The Universal Language

of

Divine Light

and

Frequencies

The Universal Theory, Volume 1

A Theory by A.J. DelVecchio

Published by A.J. DelVecchio, 2024

THE UNIVERSAL LANGUAGE OF DIVINE LIGHT AND FREQUENCIES

First edition. October 11, 2024.

Copyright © 2024 A.J. DelVecchio.

ISBN: 979-8227333681

Written by A.J. DelVecchio.

Table of Contents

Introduction: The Ancient Message from The Ancient Messengers..... 1

Chapter 1: No, You Beam Down To Me, Scotty.................................5

Chapter 2: Void the Void, on Medium Difficulty.........................27

Chapter 3: And Let There Be Divine Light with Reason 37

Chapter 4: Oh Pharaoh Tesla Stop Eavesdropping...........................71

Chapter 5: You Can Hear Voices, But Don't Worry You Can Take Meditations For That 87

Chapter 6: You're Rather Bright, God Must Say 103

Chapter 7: The Divine Medium and the Evolution of Cosmic Communication 141

Chapter 8: Quantum Communication: Because Yelling at the Universe Takes Too Long 161

Chapter 9: Divine Darkness Gravitates Around All of Us 173

Conclusion 199

Bibliography... 203

Illustrations 233

A Quote From the Author

"Similar to the universe that is everything, to black holes, the formation of stars, the celestial bodies within a star's solar system, the galaxies that surround us, and the one we reside in...time is neither linear nor unidirectional. It is cyclical within a medium that acts as a connecting web, therefore, so is our history that is bound by it. Alas, points upon this cycle can be revisited infinitely, never truly binding us existentially to time itself."
A.J. DelVecchio

Dedication

To my mother, my father, my brother, my sister, my wife, my children, and the loved ones that shared life with me throughout my time: Time may seem to pass, but the individuals that I have loved are immortalized through the medium, themselves, and myself within the time with which we spent & will spend together.

Love,
A.J. DelVecchio

Introduction: The Ancient Message from The Ancient Messengers

In the information era, the world is enveloped in information—whether invisible through technology like Wi-Fi and cellular service, or written down in digital formats and tangible books. But information is something that the human mind and soul instinctively desires—for answers, for truths, for enlightenment, and to progress throughout life. This pursuit dates back to our earliest ancestors, 200,000 years ago when homosapiens first emerged on Earth seemingly out of nowhere and unlike anything already here. Since then, human beings were always intrigued innately by information. Today, when one thinks of information, our minds usually gravitate to computation, data, and occasionally books. But most do not think about where the real information comes from and how it gets to us. But interestingly, most can recognize if it does. When you think about how information was shared and translated over 100 years ago, it required someone to take a pen, mark a piece of paper with ink, and write down their information through written language. That information needed to be either published and printed in a book to be purchased by some of the population, a note hand-delivered by the messenger, or sent in a letter to the individual that wanted to receive the information. Now, information can be delivered through the air, simply, easily, and quickly - so easily no one ever thinks about it, so simply yet nobody understands how it is done, and so quickly everyone complains when it doesn't occur within seconds. As current technological advancements have progressed, as well as advancements that humanity has perceived and ultimately achieved, it's easy to envision how technology can change everybody's lives in the future when it comes to translation and deliverance of information. Yet, what many don't realize, or have never been taught, is that information has always been transmitted far more quickly, invisibly, and efficiently throughout humanity's entire existence. For 200,000 years, humans have been able to speak to each other in ways that they did not need ink, they did not need a piece of paper, they didn't need a stamp, they didn't need a cell phone, nor cell towers, or WiFi to communicate with each other. There is a field all around us—a medium—in which information, data, light, sound, particles, molecules, frequencies, and waves influence the world we see, hear, and feel. It influences all stimuli, all matter, all energy, and is the fabric that binds and connects us to existence both here and through the cosmos. This has been happening for a very long time, and it's time everyone begins to understand the medium within reality.

Even though the following words may seem odd, I suggest reading or listening to the entirety of the book. There are many sections buried within this text that will help you as the reader live a better life, and to inform you with the proper knowledge to defend against, and defeat, those that try to hinder not only your own but your future generation's opportunities, experiences, and overall well-being. It is broken up in sections to make reading quicker and easier to review for later dates. Let us begin...

Chapter 1: No, You Beam Down To Me, Scotty

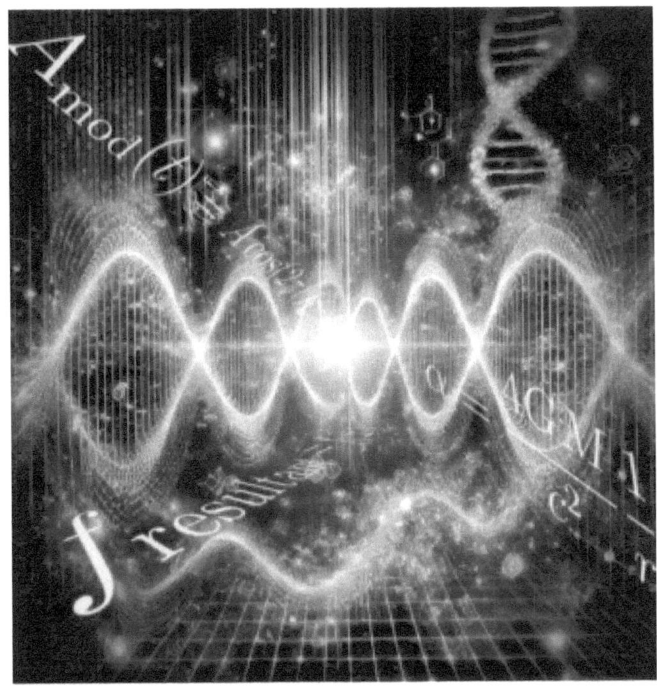

Since the dawn of human consciousness, light has been the primary medium through which the universe is interpreted and experienced. Eyes require light to see objects in the world around us. The cones within an eye absorb reflected light and detect colors based on which part of the light spectrum is not absorbed. However, the role of light extends far beyond mere perception. The human body as a whole interacts with light in important and seemingly mysterious ways—ways that can receive information to influence and evolve your physiology and psychology.

Everyone knows about getting tan or sunburned from exposure to sunlight, but this is not the data that we're discussing. If we were able to perceive all forms of light across the entire spectrum, we would witness a spectacular display of photons whizzing by, almost like shooting stars—streaks of rays producing layered lines that change what we are looking at depending on the wavelength and type of light hitting our eyes. However, the invisible light and rays that we cannot see with the cones in our eyes—the non-visible light spectrum—carry energies and information that affect the world around us on a deeper level than one might think. Humans only perceive 0.0035% of the overall frequencies and light in our world, that means 99.9965% of the surrounding universe and world cannot be experienced visually.

Light carries with it the potential to transmit vast amounts of information across unimaginable distances. As it travels from distant stars, it is not just a wave or a particle; it is a messenger, carrying the encoded information of its origin, the journey it has undertaken, and the secrets it beholds.

The Physics of Light Modulation and Information Encoding

Before we begin our journey we must review a few equations and what their symbols mean to fully comprehend my theory. It is basic physics material that I ensure will not be difficult to understand.

The ability of light to carry information is rooted in its fundamental properties. As a wave, light can be modulated in various ways to encode data. This process involves altering specific aspects of the light wave, such as its amplitude, frequency, or phase, to represent different pieces of information. This principle is not limited to visible light alone but extends across the entire electromagnetic spectrum, including radio waves, microwaves, and even gamma rays. Light exhibits both wave-like and particle-like properties, a concept described by quantum mechanics. The energy of a photon (a particle of light) is given by:

$$E = h\nu$$

Where:

- E represents the energy of the photon

- h represents Planck's constant which is

$6.626 \times 10^{-34} \text{J} \times \text{s}$

- J "Joules" amount of energy transferred

  - S represents seconds

- ν represents the frequency of the light wave

Modulation Techniques

Modulation involves varying one or more properties of the carrier wave (in this case, light) to encode information. The most common types of modulation include:

Amplitude Modulation (AM)

The amplitude (intensity) of the light wave is varied in proportion to the information being sent. It is expressed mathematically as:

$$E(t) = [1 + m(t)] \times E_0 \times \cos(\omega t)$$

Where:

- $m(t)$ represents the modulation signal (representing the information)

- E_0 represents the amplitude of the carrier wave

➢ ω (Omega) represents the angular frequency.

Frequency Modulation (FM)

The frequency of the light wave is varied according to the information being sent. It is expressed mathematically as:

$$E(t) = E_0 \times \cos(\Omega_c t + \Delta\omega \times m(t))$$

Where:

➢ $\Omega_c t$ represents the carrier frequency (Ω is Omega—but variable angular frequency)

➢ $\Delta\omega$ (Delta × Omega) represents the frequency deviation proportional to the information signal which is $m(t)$. (Omega lowercase is the base angular frequency)

Phase Modulation (PM)

The phase of the light wave is varied according to the information being sent. It can be expressed mathematically as:

$$E(t) = E_0 \times \cos(\omega t + \Delta\phi \times m(t))$$

Where:

➢ $\Delta\phi$ represents the phase deviation proportional to the information signal, $m(t)$

➢ ϕ represents Phi, or the instantaneous phase of the wave

Encoding Across the Electromagnetic Spectrum

Just as information can be encoded onto light, overall data can also be encoded onto other forms of electromagnetic radiation, such as Wi-Fi signals, FM and AM radio transmissions, and pulse modulated microwaves. Each of these wavelengths can carry vast amounts of information, with the encoding process relying on principles like amplitude modulation (AM), frequency modulation (FM), and phase modulation (PM). Wi-Fi operates in the radio frequency range, typically around 2.4 GHz and 5 GHz. These frequencies are modulated to encode and transmit data wirelessly, enabling communication between devices at the speed of light. While Wi-Fi signals may slow slightly when passing through materials, their fundamental speed in a vacuum is the same as that of visible light.

Pulse Modulated Microwaves, Gamma Waves, Frequencies, and High-Energy Radiation

Microwaves, used in radar and some communication systems, can also be modulated to carry data. Pulsed microwaves are particularly useful in applications like radar, where the time delay of the returning pulse provides information about the location and speed of objects. Gamma rays, being the highest-energy form of electromagnetic radiation, also travel at the speed of light. The energy associated with gamma rays (γ) is significantly higher, given by:

$$E\gamma = hf$$

where f for gamma rays is typically much higher than for visible light, leading to energies that can influence and change matter at the atomic level. This high energy allows gamma rays to carry information over vast distances with minimal loss of integrity, making them of interest in theoretical secure communication methods across space.

Gamma rays are a type of high-frequency electromagnetic radiation, and the equation $E\gamma = hf$ directly applies to them in the same way it does to other forms of electromagnetic radiation. Gamma rays have extremely high frequencies, typically ranging from 10^{19} to 10^{24} Hz or more. According to the equation , since f is very high for gamma rays, their photon energy $(E\gamma)$ is correspondingly very large. Because of their high frequency, gamma rays have much higher photon energy compared to other electromagnetic waves like visible light or X-rays. This high energy makes gamma rays capable of penetrating materials and causing significant interactions with matter, such as ionizing atoms and molecules. In contexts such as nuclear reactions or particle physics, gamma rays are often produced when an atomic nucleus undergoes radioactive decay. The energy released in these processes is carried away by photons, and the energy of these gamma-ray photons is calculated using the same equation.

For example, if a gamma-ray has a frequency of 5×10^{20} Hz, its energy would be:

$$E_\gamma = hf = (6.626 \times 10^{-34}) \times (5 \times 10^{20}) = 3.313 \times 10^{-13} J$$

$$\text{or } 0.0000000000003313 J$$

This energy is very large for a single photon, which is why gamma rays are so powerful and used in applications such as medical imaging, cancer treatment, and sterilization, as well as in astrophysical observations where they are emitted by high-energy cosmic events like supernovae or black hole activity.

The Universality of Light Speed in Communication

One of the unifying principles across all these different types of electromagnetic radiation is that they all travel at the speed of light in a vacuum. This speed—299,792 kilometers per second (or about 186,282 miles per second)—is a constant, regardless of the frequency or wavelength of the radiation. Whether it be a Wi-Fi signal, an FM radio broadcast, or simply a pulse of microwaves being transmitted, the information encoded on these waves reaches its destination at the same speed. Objects within the world can prohibit specific frequencies from reaching their destination. Most light can be blocked by matter in the universe, but many types of frequencies are not blocked by matter at all, whizzing through objects with ease.

Mathematics of Light Speed and Data Transmission

The speed of light C is a fundamental constant in physics appearing in equations that we all recognize, such as the energy-mass equivalence

equation:

$$E = mc^2$$

The wave equation is:

$$c = \lambda f$$

Where:

- λ is called Lambda or the wavelength

- f is the frequency

This equation illustrates the inverse relationship between the frequency and wavelength of light or other electromagnetic waves. Higher frequency waves, such as gamma rays, have shorter wavelengths, while lower frequency waves, such as radio waves, have longer wavelengths. λ is defined as the distance between consecutive points of the same phase on a wave, such as from crest to crest or trough to trough. It is a key parameter in describing waves, including light, sound, and other electromagnetic waves. Wavelength is related to the speed of the wave and its frequency by the equation:

$$\lambda = v/f$$

Where:

➤ λ is the wavelength

➤ v is the wave's velocity in meters per second

➤ f is the frequency of the wave in hertz (Hz).

➤ A is the Amplitude of the wave in the figure.

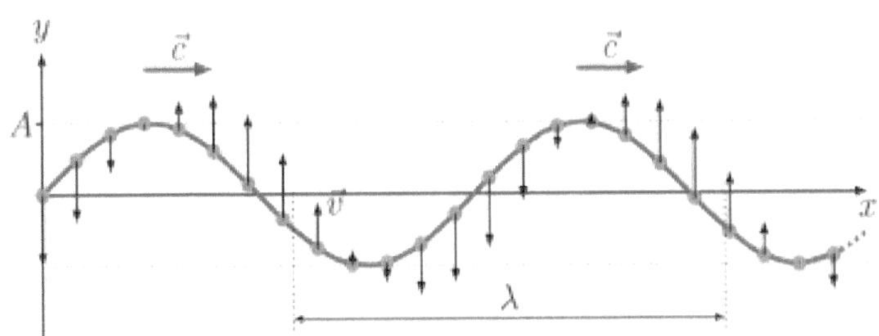

How Terrorists and Evil People of the World Fake Being God

The patent US3951134A outlines a method and apparatus for remotely monitoring and altering brain waves. This patent is from 1975-1976, about 50 years old, during a time when only FM and AM radio frequencies bathed the medium around us. Today, we have many different frequencies surrounding us at all times, with antenna and frequency emitters in our pockets. Considering current technological advancements, and consistent bathing in 5G frequencies, it would be very simple for anyone to perform methodologies described in this patent. It's a fascinating and, admittedly, slightly unnerving process that works like this:

To begin, typically two distinct electromagnetic signals, such as 100 MHz and 210 MHz, are transmitted toward the subject's brain. These signals penetrate the skull and interact within the brain tissue, producing an interference pattern. This interference pattern is similar to how binaural beats are generated and created, which is simply two different frequencies played in each ear. You can actually try this yourself with headphones and a website that plays pure tone frequencies. For example, the frequency in the left ear is 100 Hz and the frequency in the right ear is 150 Hz. The resulting new frequency between the two that the brain hears is 125 Hz. You can still hear 100 Hz in one ear and 150 Hz in the other, but there is a third frequency of 125 Hz that resides ominously between your ears. (We will discuss binaural beats, and which frequencies play a role in the brain's function in a later chapter. For now, let's continue the discussion about how targeted individuals are harassed by similar methods.) It's not just about bombarding the brain with frequencies; these waves combine, and each region of the brain uniquely modulates the interference pattern based on its natural electrical activity. The modulated interference pattern, now carrying the brain's electrical signature, is retransmitted back through the skull. Essentially, the brain becomes a transmitter, broadcasting a signal containing information about its internal workings. This signal is then captured by a remote receiver, where it is demodulated to extract the detailed profile of the brain's electrical activity. This entire process occurs from a distance without the need for physical contact, making it possible to monitor, alter, and influence the brain's activity in real-time. Once received, the signal undergoes further processing to isolate the specific components representing brain wave activity. A mixing process is used to extract a signal frequency corresponding to the difference between the two transmitted frequencies. This mixing process is similar to a conventional frequency mixing technique used in radio communications. Each region of the brain modulates these signals based on its electrical activity, which alters the resulting interference pattern. This resultant signal, modulated by the brain's natural activity, provides a clear, real-time representation of the brain's internal state. What makes this technology even more compelling is its bidirectional capability. Beyond just monitoring, the system can generate compensating signals based on the received brain wave data and transmit these back to the brain. This creates a feedback loop, allowing for the precise modification of the brain's electrical activity, influencing neurological processes in a targeted manner. This means the subject, or a targeted individual, can experience complete thoughts, auditory experiences, and motor movements that are instructed by an external input. Essentially, a person's bodily movements and thoughts can be easily choreographed by someone with the right technology.

The underlying method relies on standard RF signal mixing and modulation principles, depicted as follows:

Signal Combination in the Brain:

$$f_{resultant} = f1 \pm f2$$

Example:

$$f_{resultant} = 210\,MHz - 100\,MHz = 110\,MHz$$

$$f_{demodulated} = f_{resultant} - f_{reference} = 10\,MHz$$

$f_{resultant}$: is the frequency resulting from the interaction of $f1$ and $f2$. It reflects either the sum or difference of these initial frequencies, indicating how the brain processes the incoming signals.

Modulation of the Resultant Signal:

$$A_{mod}(t) = A\cos(2\pi(f1 - f2)t + \phi(t))$$

$\phi(t)$: Denotes the phase modulation component influenced by the brain's natural electrical signals.

This equation illustrates how the brain modifies the amplitude and phase of the combined signal, essentially encoding its own electrical activity onto the incoming waves.

Demodulation Process:

$$A(t) = \text{Demod}\left[A_{\text{mod}}(t)\right]$$

The process described in the patent involves receiving a modulated signal (resulting from the interaction of two transmitted frequencies within the brain) and demodulating it to extract information about brain activity. Here's a step-by-step explanation of the demodulation process, along with the associated mathematical expressions.

Received Signal:

The received signal, $S_{\text{received}}(t)$, after being modulated by brain wave activity, is described as:

$$S_{\text{received}}(t) = A_{\text{mod}}(t)\cos\left(2\pi f_{\text{resultant}}(t) + \phi(t)\right)$$

Where:

➤ $A_{\text{mod}}(t)$ is the amplitude modulation due to brain waves.

➤ $f_{\text{resultant}}$ is the resultant frequency of the two mixed signals (e.g., 110 MHz).

➤ $\phi(t)$ is the phase modulation, if present, due to brain wave activity.

Reference Signal:

A reference signal, $S_{reference}(t)$, of frequency $f_{reference}$ (e.g., 100 MHz), is used in the demodulation process:

$$S_{reference}(t) = \cos(2\pi f_{reference}\, t)$$

Mixing Process:

The received signal is mixed with the reference signal to produce sum and difference frequencies. The resulting mixed signal, $S_{mixed}(t)$, is given by:

$$S_{mixed}(t) = S_{received}(t) \times S_{reference}(t)$$

If we substitute the expressions for $S_{received}(t)$ and $S_{reference}(t)$

we get:

$$S_{mixed}(t) = A_{mod}(t)\cos(2\pi f_{resultant}(t) + \phi(t))\cos(2\pi f_{reference}t)$$

Using the trigonometric identity:

$$\cos A \cos B = \tfrac{1}{2}\left[\cos(A - B) + \cos(A + B)\right]$$

We get:

$$S_{mixed}(t) = (A_{mod}(t)/2)\left[\cos(2\pi(f_{resultant} - f_{reference})t + \phi(t)) + \right.$$

$$\left. \cos(2\pi(f_{resultant} + f_{reference})t + \phi(t))\right]$$

Filtering Process:

The mixed signal contains two components:

Difference Frequency:

$$f_{difference} = f_{resultant} - f_{reference}$$

Sum Frequency:

$$f_{sum} = f_{resultant} + f_{reference}$$

The difference frequency contains the desired demodulated signal, while the sum frequency is discarded using a band-pass filter.

$$S_{filtered}(t) = (A_{mod}(t)/2)\cos(2\pi(f_{resultant} - f_{reference})t + \phi(t))$$

For the example in the patent:

$$f_{difference} = 110\,MHz - 100\,MHz = 10\,MHz$$

The filtered signal becomes:

$$S_{filtered}(t) = (A_{mod}(t)/2)\cos(2\pi \times 10\,MHz \times t + \phi(t))$$

Demodulation to Extract Brain Activity:

Depending on the modulation type (amplitude, frequency, or phase), different techniques are used to extract $A_{mod}(t)$ and/or $\phi(t)$:

Amplitude Demodulation extracts $A_{mod}(t)$ by using an envelope detector:

$$A(t) = \left| S_{filtered}(t) \right|$$

Frequency Demodulation measures the instantaneous frequency:

$$f_{instantaneous} = 1/2\pi \times d/dt \left[2\pi(f_{difference}t + \phi(t)) \right] =$$

$$f_{difference} + 1/2\pi \times d\phi(t)/dt$$

Phase Demodulation extracts $\phi(t)$ directly by using a phase-locked loop (PLL) or other phase demodulation techniques.

Resulting Demodulated Signal:

The final demodulated signal, which corresponds to the brain's electrical activity, can be represented as:

$$S_{demodulated}(t) = A_{mod}(t) \text{ or } \phi(t)$$

Where:

➤ $A_{\text{mod}}(t)$ represents the amplitude variations.

➤ $\phi(t)$ represents the phase variations, depending on the type of modulation used by the brain wave activity.

This operation pulls the modulated signal apart, extracting the brain's natural electrical activity from the received signal.

These equations showcase how the process of signal transmission, interference, modulation, and demodulation works together to create a clear picture of the brain's electrical activity. It's a two-way interaction where the brain acts both as a receiver and a transmitter, with external signals altering its internal processes.

Gravitational Lensing Effect

Another crucial aspect is how light and electromagnetic waves can be influenced by massive objects through gravitational lensing. According to Einstein's General Theory of Relativity, light is bent when it passes near a massive object due to the curvature of spacetime. The angle of deflection for light passing close to a massive object, such as a star, can be described by:

$$\alpha = 4GM/c^2 \times 1/r$$

Where:

➤ α is the angle of deflection

➤ G is the gravitational constant

➤ M is the mass of the lensing object

➤ c is the speed of light

➤ r is the distance of the closest approach of the

light to the object.

This gravitational lensing can focus and magnify light and other electromagnetic signals, potentially carrying encoded information across vast distances and focusing it onto specific points, such as Earth. Given that all forms of electromagnetic radiation travel at the same speed in a vacuum, this principle supports the idea that information carried by light from distant stars can be transmitted across the universe. Just as Wi-Fi or radio waves are utilized to transmit information across the Earth, the cosmos might be using light and other forms of radiation to transmit information across the vastness of space. The gravitational lensing effect could focus and magnify this information, allowing it to reach Earth more effectively and potentially influence life here in ways science and physics is only beginning to understand.

The Interplay of Light and Human Biology

Beyond technological and cosmic implications, light interacts with human biology in intricate ways. The human body is not just a passive receiver of light; it actively engages with it, absorbing and using light in various physiological processes. Light, particularly in the form of sunlight, triggers various biological processes in the human body. Certain wavelengths of light can influence mood, circadian rhythms, and even cellular processes. These interactions underscore the idea that light carries information that can affect biological systems.

While visible light is crucial for vision and many biological processes, non-visible light, such as ultraviolet (UV), infrared (IR), and other forms of electromagnetic radiation, plays equally significant roles in human health and the environment. UV light, for instance, is involved in the synthesis of vitamin D in the skin but can also cause DNA damage, leading to skin cancer. The interaction of these different types of light with the human body suggests that light carries more than just visual information; it influences our biological functions and can even affect our mental and emotional states. This further strengthens the idea that light, in all its forms, is a carrier of information that has shaped and continues to shape life on Earth. Infrared light, on the other hand, is perceived as heat and is essential for thermal regulation. Red and near infrared light is capable of penetrating the skin, where it is absorbed by the cell's main powerhouse: mitochondria. Cytochrome C-Oxidase in mitochondria is activated, particularly in the range of 600-900 nm. This stimulation initiates cellular regeneration throughout the body. It also helps improve cellular energy production called Adenosine Triphosphate (ATP) synthesis, fat metabolization, while reducing inflammation overall. ATP is produced in the mitochondria through a process called cellular respiration, specifically in the stages of glycolysis, the citric acid cycle, and oxidative phosphorylation. The bodily process used for wound healing, pain reduction, tissue repair, improving skin health, lowering oxidative stress, and reducing wrinkles is called Photobiomodulation (PBM).

Recent studies suggest that blue light (short wavelengths between 400-500 nm) can influence DNA repair by regulating the circadian clock. Circadian rhythms are the natural, internal processes that regulate the sleep-wake cycle and repeat roughly every 24 hours. These rhythms are driven by the body's biological clock and influence many bodily functions, including hormone release, digestion, and body temperature. The most well-known circadian rhythm is the sleep-wake cycle, which is influenced by external cues such as light and darkness. Suprachiasmatic Nucleus (SCN) is a region in the hypothalamus of the brain that acts as the master clock, regulating circadian rhythms by responding to light exposure. A hormone that helps regulate sleep, produced in greater amounts when it's dark, signaling the body to prepare for sleep. Light plays a critical role in aligning the circadian rhythms with the environment, especially blue light, which helps signal the brain when to wake up.

When circadian rhythms are disrupted—due to factors such as jet lag, shift work, poor sleeping habits or excessive exposure to artificial light—people may experience sleep disorders, mood disturbances, and long-term health issues such as increased risk for cardiovascular disease and metabolic disorders. Cells repair DNA damage more efficiently during certain times of the day, a process governed by circadian rhythms. Blue light exposure can help entrain the circadian clock, ensuring that DNA repair processes occur at their optimal times. Proper circadian alignment is crucial for the efficiency of DNA repair. Disruption of circadian rhythms, such as through excessive exposure to artificial light at night, can impair the body's natural ability to repair DNA, increasing the risk of developing cancer and other diseases. On the other hand, carefully timed blue light exposure can enhance DNA repair by maintaining a healthy circadian rhythm.

As of this writing, humanity has the capability to sequence nearly the entire human genome or DNA. However, our functional understanding is limited to only 1-2% of the genome, which directly codes for proteins. The other 98% is referred to as "non-coding DNA." This portion of the genome plays crucial roles in regulatory functions, gene expression, and other complex processes that science and medicine do not yet fully understand. While humans may grasp how light can affect a cell, our comprehension of how light and frequencies interact with our DNA remains limited. Science knows that UV, near-infrared, blue light, and other light spectrums can activate specific cellular processes, such as DNA repair and circadian rhythm regulation. However, given that 98% of human DNA remains uncharted territory, it is plausible that light and frequencies could trigger processes encoded within our DNA, potentially leading to evolutionary changes in humanity or certain individuals. As science delves deeper into the mysteries of non-coding DNA, humans may uncover pathways that are currently beyond our understanding.

The idea that our DNA is packed with sequences that can be triggered or modified through light and frequencies, one could argue that this is part of the divine design of human beings. Personality, physical, mental, emotional, and spiritual states reside in the DNA of every human, waiting for the proper communication from the divine. It may not be within our immediate intellectual grasp to fully comprehend how God uses these mechanisms to guide human evolution. Nonetheless, the pursuit of knowledge in this area is crucial. If humanity aims to evolve towards a Type V civilization—capable of mastering universal scale energy and knowledge—humans must first progress through Type I to IV on the Kardashev Scale, which involves harnessing planetary, stellar, and galactic energy resources, respectively (we will discuss this scale in a later chapter). Failure to achieve these milestones could risk our self-destruction, illustrating the delicate balance between human ambition and the need for guided progression. This need for guided progression seems to be a necessity for advancing into Type I and beyond, unless humans can mimic God's duties and gene evolution to a point where they can dissolve all evils that hinder them as a species from reaching, at the minimum, Type I. My suggestion is that humanity actively pursues the comprehension of God's influence and divine intervention, while maintaining a strong connection to the divine. This would provide humanity with a safety net while progressing and evolving as a species.

The Potential of Light for Future Communication Technologies

As science advances in the understanding of light and its properties, the potential applications for light in communication technology continue to grow rapidly. Technologies such as Li-Fi (Light Fidelity) are emerging, which use light waves instead of radio waves to transmit data. Unlike Wi-Fi, which relies on radio frequencies, Li-Fi uses the visible light spectrum to deliver high-speed data transmission. This technology could revolutionize the way the world accesses the internet, providing faster and more secure connections. In the future, the world may develop even more advanced methods of encoding and decoding information carried by light, potentially allowing humans to communicate across vast distances, even interstellar ones. These technologies could open up new possibilities for exploring the universe and perhaps even detecting signals from other civilizations.

Now that we are finished reviewing some basic knowledge necessary to comprehend my theory, let us begin the ongoing—even after this book is published and *you* are reading it—discussion about how there is a universal language through light and frequencies provided by the divine that touch our souls, and aid everyone in transcending into and traversing the medium in which we all reside. Enjoy.

Notes

25

Chapter 2: Void the Void, on Medium Difficulty

Light, often taken for granted in our everyday experience, plays a much deeper role in the fabric of reality than most people realize. From the beginning of time, light has been the universal medium through which the cosmos communicates with itself. It's not just a means of seeing but a conveyor of vast amounts of information that has truly traveled across unimaginable distances. This chapter explores how light is not merely an illuminating force but a cosmic messenger carrying the secrets of the universe.

Light and the Human Experience

Humans have always been drawn to light, from the ancient rituals revolving around the sun to the awe inspired by the night sky. But what if the light everyone visually sees is just a fraction of what truly exists? Beyond the visible spectrum lies a vast array of electromagnetic waves, each carrying its unique form of energy and information. This light, both visible and invisible, interacts with our bodies, influencing us in ways that science is only beginning to understand.

Light affects all humans biologically, psychologically, and spiritually. For instance, the circadian rhythms that govern our sleep-wake cycles are directly influenced by light, particularly the blue wavelengths. But beyond some of these well-known effects, light in all forms also carries information that can influence our cells at a more distinctive level. As we delve deeper into the science of light, we find that it may play a role in the transmission of information between local and distant stars and our planet, impacting and influencing everything from the development of life to the evolution of consciousness. If any intelligence similar or more advanced than us exists within the cosmos, then the usage of light to transmit information and data to alter and influence the human species cannot be dismissed.

The Physics of Light Transmission

To understand how light can carry information, we must delve into the physics of light itself. Light behaves both as a wave and a particle, a duality that allows it to interact with the universe in complex ways. When light waves are modulated—by altering their amplitude, frequency, or phase—they can encode data, much like how radio waves carry information to our devices. This is the principle behind technologies like fiber optics and Li-Fi, which use light to transmit data or information at incredible speeds.

But light from distant stars also carries natural modulations, encoded with data about the star's composition, movement, and interactions with surrounding cosmic matter. As this light travels across the universe, it retains this information, arriving at Earth as a message from the stars. This encoded light could potentially interact with our own energy fields, influencing our physical and spiritual states.

The Potential of Light in Information Transmission

The potential for light to carry information extends far beyond our current technological capabilities. While science and engineering has developed methods to transmit data using visible light and lasers, the universe has been doing this on a cosmic scale for billions of years. The light that reaches Earth from distant galaxies has traveled across vast stretches of space-time, bringing with it information that could be crucial to the understanding of the universe.

Quantum entanglement further complicates the understanding of how information is transmitted through light. When two particles become entangled, their states are linked, no matter how far apart they are. This phenomenon suggests that information could be transmitted instantaneously across vast distances, a concept that challenges our traditional understanding of light and communication.

The Role of Non-Visible Light

While visible light is crucial to everyone's daily existence, non-visible light plays an equally important role in the transmission of information. Infrared, ultraviolet, X-rays, and gamma rays all carry different types of information, interacting with matter in unique ways. These forms of light can penetrate objects, reveal hidden structures, and even alter the properties of the materials they interact with.

For example, gamma rays, which are the highest energy form of light, can be used to study the most energetic and extreme events in the universe, such as supernovae and black holes. These rays carry information about the most violent processes in the cosmos, potentially influencing the very fabric of space-time. As humanity continues to explore the universe, understanding the role of non-visible light will be key to unlocking new secrets about the nature of reality.

The Medium of Light and Quantum Information

The concept of light as a medium of information is deeply connected to the idea of quantum mechanics. At the quantum level, particles such as photons (the particles of light) exhibit behaviors that challenge our classical understanding of the universe. Quantum entanglement, for instance, suggests that particles can be linked across vast distances, allowing for the instantaneous transfer of information—a phenomenon that could potentially explain how light from distant stars carries specific data to Earth.

Moreover, the concept of the quantum field, a fundamental framework in physics, supports the idea that light is part of a larger, interconnected system where information is constantly being exchanged. This field is not bound by traditional notions of space and time, which means that the information carried by light could interact with other aspects of the universe in ways science is only beginning to comprehend.

The Application of Light in Modern Technology

Modern technology has only begun to tap into the potential of light for information transmission. Fiber optics, for instance, revolutionized telecommunications by using light to transmit data at speeds far greater than traditional methods. Li-Fi, a wireless communication technology that uses light to transmit data, offers even greater possibilities, including the potential for faster and more secure communications.

But the most intriguing applications may lie in the realm of quantum communication. Quantum computers, which leverage the principles of quantum mechanics, are expected to revolutionize our ability to process information. These computers use qubits, which can exist in multiple states simultaneously, allowing for parallel processing on a scale that classical computers cannot achieve. Light, with its quantum properties, could play a central role in the development of these technologies, enabling the transmission of quantum information across the globe

Unknown Modulated Frequencies from the Cosmos

There have been several one-time modulations in frequencies emitted from the cosmos and recorded by scientists worldwide. These events don't necessarily fit neatly into the patterns associated with known phenomena like pulsars, fast radio bursts, or supernovae. All of the signals listed below have been detected, recorded, and never observed again, making them one-time modulations of frequencies that have reached Earth. This makes them very difficult to study, but raises an intriguing point about how frequencies and light can reach our planet while being modulated by unknown forces.

The "Wow!" Signal of 1977

The most famous signal received on Earth that was modulated in such an odd way it did not align with known physics or observations of star-emitting objects in the cosmos is the WOW signal from 1977. This signal lasted for 72 seconds and has never been detected again. It was a very strong, narrowband radio signal observed in the 1.42 gigahertz range, a frequency close to that of neutral hydrogen, which science can only partially explain. It was detected by astronomer Jerry Ehman, and despite repeated searches, it has never been observed again. Even to this day, scientists cannot definitively explain its origin. It does not point directly to a supernova, black hole, or any other known star system.

It is particularly important because it was detected at the frequency naturally emitted by neutral hydrogen, the most abundant element in the cosmos. This frequency is ideal for interstellar communication, as signals at this wavelength can travel vast distances without being disrupted or experiencing decoherence by the background noise of the interstellar medium or other objects crossing its path.

The signal lasted only 72 seconds, which was the maximum duration the Big Ear radio telescope could record at the time due to Earth's rotation. Humanity is unsure if it lasted longer or if it changed after the recording made by Jerry Ehman. Given its extremely narrow bandwidth, this signal stands out because almost all natural sources of radio waves, such as stars, galaxies, pulsars, supernovae, and black holes, and magnetars typically emit across a much broader range of frequencies or altogether differently than what was observed. This suggests the "Wow!" signal could be artificial in origin.

The signal was 30 times stronger than the background noise, and considering its narrow bandwidth, strength, and duration, it is unlikely to have come from a natural source close enough to emit that type of frequency. While no simple message was decoded from the signal (such as one written in English or another language), do not rule out the possibility that the signal was encoded with information and modulated to convey data, or even to potentially deliver biologically altering instructions for us and the worlds we all reside in. It's also possible that the "Wow!" signal was part of a more complex transmission that either didn't repeat or didn't last long enough to reveal its full message. Earth's rotation is expressed in the released observation, showing that the rotation faded the signal in and out of the Big Ear Radio telescope. There could have been other methods used to encode the frequency, like phase modulation (PM) or spread-spectrum techniques, which would require advanced detection methods not available at the time. Even if the telescope had the advanced technology needed, would humanity have had the knowledge at the time to decode the message properly? Most likely not. One spread-spectrum technique that would work and go unobserved is Frequency Hopping Spread Spectrum (FHSS), where the carrier frequency of the signal is rapidly switched (hopped) between different frequencies within a specified range. The hopping pattern follows a predefined and somewhat random sequence, and both the transmitter and receiver must know the sequence in advance to communicate effectively. This points towards direct biological changes that only DNA or other aspects of life would receive and respond to accordingly. This aspect is spoken about in later chapters, so do not fret. Simply store this part in the back of your mind for when you reach that part of the book. While hydrogen clouds can emit frequencies in the 1.42 GHz range, they are almost always broad and continuous, whereas the "Wow!" signal was narrow and transient. Furthermore, the "Wow!" signal didn't originate from Earth, as it was detected within a protected frequency band not allowed for terrestrial transmissions, making Earth-based explanations impossible. It was tied to the hydrogen line of potential waves, a highly logical choice for anyone attempting to communicate across vast cosmic distances. Despite almost 50 years of observation, no similar signals have been detected, and none have been observed in the same region of the sky.

The Implications for Human Evolution

As we continue to explore the possibilities of light as a medium of information, the implications for human evolution are extensive. The ability to harness light for communication, computation, and understanding could lead to a new era of human development, one where our connection to the cosmos is not just metaphorical but literal. The ancient wisdom that has long recognized the significance of light—whether through religious texts, spiritual practices, or early scientific exploration—may hold clues to our future. By studying the role of light in the universe, scientific discoveries may unlock new pathways for evolution, both on a personal and a species-wide level. This journey of understanding light as a universal language could bring us closer to a deeper understanding of our place in the cosmos and the potential that lies within us.

Notes

Chapter 3: And Let There Be Divine Light with Reason

The birth and life of Jesus Christ have been central to the development of Christianity, shaping the spiritual beliefs of billions. Yet, beyond the religious narratives, there exists a possibility that his conception was not merely a miraculous event but one influenced by the cosmos itself. Consider this: Jesus was conceived between March 20th and April 25th a period marked by the alignment of stars within the Pisces constellation behind the Sun, it opens the door to an intriguing exploration of how celestial events, light from another star system, and the intelligence that has encoded it with vast information may have played a role in his divine and immaculate conception.

The Timing of Jesus' Conception

Traditionally, Jesus' birth is celebrated on December 25th, which implies that his conception would have occurred approximately nine months earlier, around March 25th. Historians claim his birthday may have occurred in spring or summer, considering the scriptures that describe shepherds catering to cattle at a specific time at night during his birth, which would refer to a warmer time of year. However, we will use his celebrated date as an example. March 20th to 21st coincides with the vernal equinox, a time of year when the Earth experiences equal amounts of daylight and darkness, symbolizing balance and renewal. It is during this time that the Sun passes in front of the constellation Pisces, which holds significant astrological and spiritual meaning. Pisces, often associated with spirituality, intuition, and compassion, is symbolized by two fish swimming in opposite directions, which represents overall duality and interconnectedness. In the context of Jesus' life, Pisces is a fitting symbol for his role as a spiritual teacher who bridged the human and divine. Between March 20th and April 25th, many stars within the Pisces constellation align behind the Sun, effectively obscuring them from view on Earth. This alignment, along with other alignments of stellar objects, is known as a syzygy, occurs when three or more celestial bodies are positioned in a straight line. As the Sun passes through Pisces, its gravitational field exerts a powerful influence on the light from these stars, causing a phenomenon known as gravitational lensing.

***Figure 1.** Depicts the sky on March 23rd, 30 B.C. This shows the alignment during the Vernal Equinox at that time. This alignment occurs during Jesus' resurrection, which is around the same date in 30-37 A.D.*

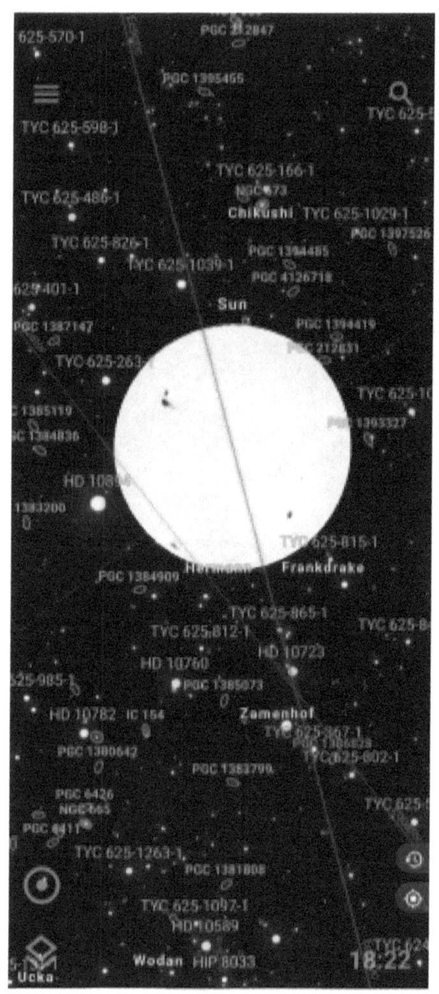

***Figure 2.** Shows just how many celestial and cosmic objects are within the Sun's range on March 23rd 30 B.C. Some are thousands of light years away, while some are only one hundred.*

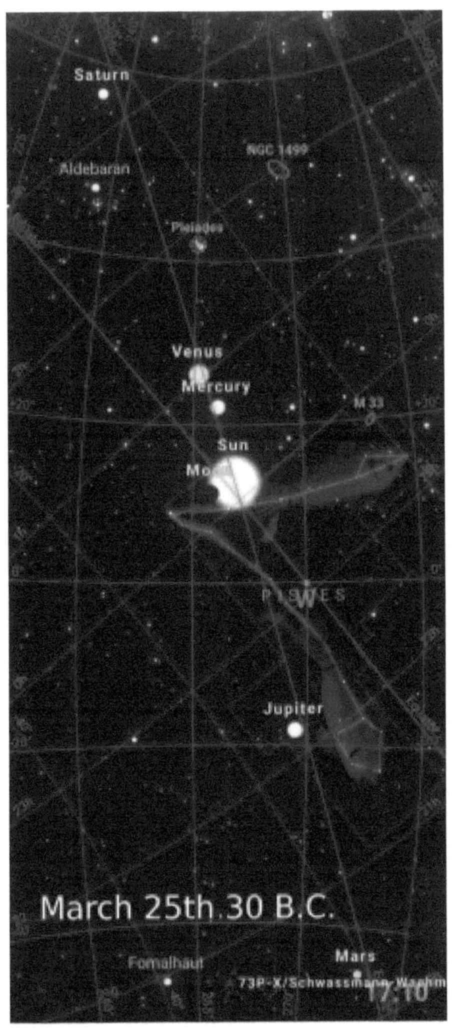

Figure 3. Shows how celestial bodies aligned within the timeframe of Jesus' conception. Even the moon makes an appearance within this timeframe. As the hours shift through the day and night from March 20th - March 25th, many cosmic and celestial objects join the Sun and Ecliptic, eventually creating a straight line. This was taken at 17:10 on March 25th, 30 B.C.

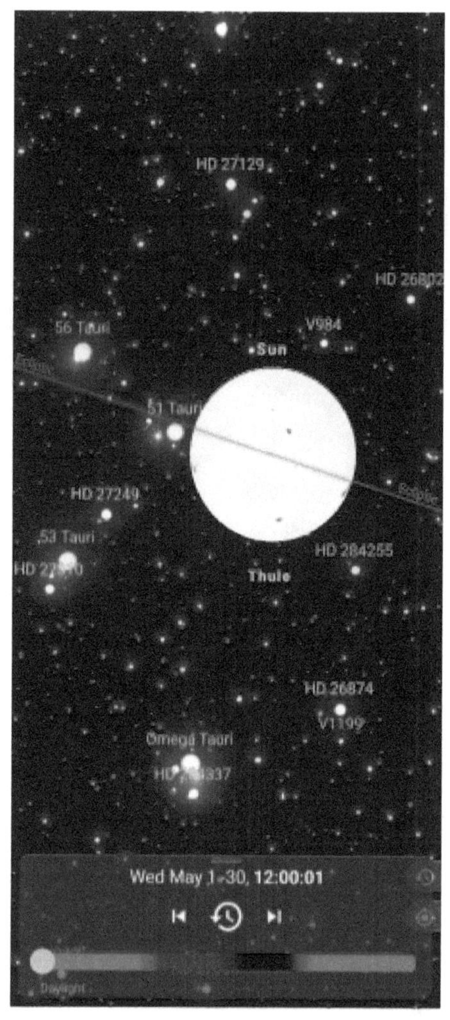

Figure 4. Depiction of the position of various cosmic and celestial objects that the sun will traverse during the time in which Jesus' first heart beat occurs.

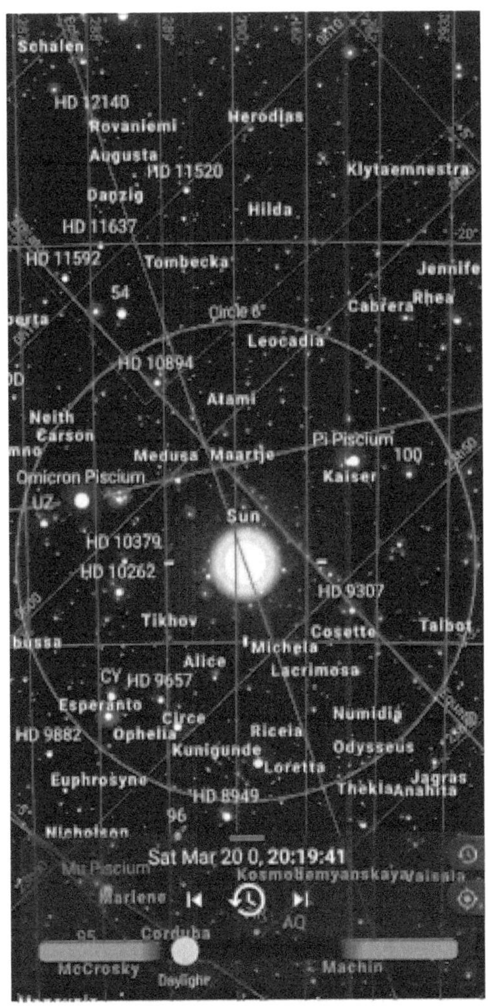

Figure 5. *This image shows us the stars and solar system objects (asteroids or minor planets) that are within the range of gravitational lensing by the sun during the beginning stages of Pisces. This includes between 466 BC and 7 AD, covering the timeline of Jesus' potential conception, first heart beat, birth, death, and resurrection. See the list below for details on all the stars that fall within this range, and the distance of the asteroids or minor planets as well. There are galaxies within this range, but all galaxies are tens to hundreds of million light years away.*

The range for the Age of Pisces and its area of precession falls between 1 AD and 2150 AD when dividing the entirety of precession into separate ages equally. However, the edge of a constellation is far from definitive and absolute. When viewing the age from the start of the constellation to the end, Pisces exists between 466 BC and 2605 AD specifically. Therefore, in *Figure 5*, I have provided a range to review. Within this range are stars that would have gravitationally lensed during this time, leading up to his resurrection. Many of these celestial bodies are a hundred or fewer light-years away, which covers the possibility that God or divine beings would have abided by the limitations of light-speed travel to transmit divine information through light and frequencies from this area of the constellation, arriving in time for Jesus' immaculate conception and miraculous resurrection.

There are also minor planets, or asteroids, in this image specifically focused on Jesus' death and resurrection. They are typically around 3+ AU (Astronomical Units) away from Earth. Every known planet becomes shrouded by the Sun many times during the full date range and a few times between conception and first heart beat, while Uranus, Saturn, Mars, the Moon, Venus, and Jupiter are all within the range of gravitational lensing between 7 BC and 7 AD (current margin of error is +/- 7 years). While there are no confirmed exoplanets in these star systems, many have not been extensively researched or even studied for exoplanets. Most of these stars are within the 480-year range, which means that if a signal was sent to Earth, it could have begun during the "earlier" stages of Pisces and arrived during Jesus' conception, first heartbeat, birth, death, and resurrection. This would, regardless, coincide with the precessional age change into Pisces. Here is a detailed list of the objects and their distances:

Stars within the Circle:

1. HD 10782

Distance: 294.36 light-years

Type: F-type main-sequence star

Significance: As an F-type star, HD 10782 is hotter and more luminous than the Sun. It is a target for potential exoplanet discovery.

Probability of Finding Exoplanets: Moderate. F-type stars are good candidates for planet searches, though they tend to have shorter lifespans than Sun-like stars, possibly reducing the stability of planetary systems.

2. HD 10262

Distance: 178.42 light-years

Type: Main-sequence star (specific type unclear)

Significance: While not extensively studied, its proximity makes it a good target for future planet-hunting missions.

Probability of Finding Exoplanets: Moderate to high due to its proximity and likely being a relatively stable main-sequence star.

3. HD 9657

Distance: Unknown (likely within 200 light-years, needs specific study)

Type: Main-sequence star

Significance: Not widely studied, but it could be a target for future searches.

Probability of Finding Exoplanets: Unknown, but likely moderate based on its spectral type and presumed proximity.

4. HD 9496

Distance: 153 light-years

Type: F-type star

Significance: F-type stars are interesting for exoplanet exploration due to their wider habitable zones.

Probability of Finding Exoplanets: Moderate. This star is close enough for detailed observation, and its type suggests it could host planets in a stable orbit.

5. HD 10894

Distance: 165 light-years

Type: F-type main-sequence star

<u>*Significance:*</u> Its proximity and spectral type make it a good candidate for future planet discovery.

<u>*Probability of Finding Exoplanets:*</u> Moderate to high.

6. HD 11520

<u>*Distance:*</u> **138 light-years**

<u>*Type:*</u> *G-type main-sequence star (similar to the Sun)*

<u>*Significance:*</u> *As a G-type star, HD 11520 is highly interesting for planet hunting. G-type stars like our Sun are more likely to host Earth-like planets in their habitable zones.*

<u>*Probability of Finding Exoplanets:*</u> *High. G-type stars are prime candidates for hosting habitable planets.*

7. HD 11637

<u>*Distance:*</u> **85 light-years**

<u>*Type:*</u> *G-type star*

<u>*Significance:*</u> *Very close and similar to the Sun, making it an excellent target for exoplanet searches.*

<u>*Probability of Finding Exoplanets:*</u> *High. Its proximity and classification suggest good potential for hosting planets.*

8. HD 9307

<u>*Distance:*</u> 155 light-years

<u>*Type:*</u> F-type star

<u>*Significance:*</u> F-type stars can have larger habitable zones than G-type stars, though their higher luminosity can affect planetary atmospheres.

<u>*Probability of Finding Exoplanets:*</u> Moderate.

9. HD 9817

<u>*Distance:*</u> 248 light-years

<u>*Type:*</u> Main-sequence star

<u>*Significance:*</u> This star hasn't been extensively studied but could be a candidate for planet searches.

<u>*Probability of Finding Exoplanets:*</u> Moderate.

10. HD 8949

<u>*Distance:*</u> **78.7 light-years**

Type: G-type star

Significance: Given its proximity and similarity to the Sun, this star is a prime target for exoplanet searches.

Probability of Finding Exoplanets: High. Being a G-type star, it has good potential for hosting habitable planets.

11. HD 9882

Distance: 187 light-years

Type: F-type star

Significance: Its proximity and spectral type make it interesting for potential exoplanet detection.

Probability of Finding Exoplanets: Moderate.

12. HD 8703

Distance: 97 light-years

Type: G-type star

Significance: Close and similar to the Sun, making it an excellent candidate for planet hunting.

Probability of Finding Exoplanets: High.

13. Pi-Piscium

Distance: 114.44 light-years

Type: Binary star system

Significance: Binary systems can host planets, but the gravitational interactions make it more difficult.

Probability of Finding Exoplanets: Moderate. While planets can exist in binary systems, their orbits tend to be more complex.

14. Omicron-Piscium

Distance: 279.5 light-years

Type: K-type giant star

Significance: As a giant star, it's less likely to host habitable planets, but studying its characteristics can help us understand stellar evolution.

Probability of Finding Exoplanets: Low. Giant stars have evolved past the stage where their habitable zones would be stable for long-term planetary life. However, this would provide motivation for an intellectual being to look for another home.

Minor Planets or Asteroids within the Circle During Jesus' Death and Resurrection:

1. Atkinson
2. Inkeroi
3. Ottilia
4. Swasey
5. Burdigala
6. Yunnan
7. Onnie
8. Van Houten
9. Bauersfelda
10. Somov
11. Rollandia
12. Beagle
13. Tanina
14. Happelia

Minor Planets could be utilized as a satellite or intermediary access point to transmit data through gravitational lensing within our own solar system.

Notable Galaxies within the Circle:

1. NGC 524

Distance: ~90 million light-years

2. NGC 489

Distance: ~97 million light-years

While faster-than-light information transmission could be plausible, if it is not possible without using an Einstein-Rosen Bridge, or wormhole, then utilizing gravitational lensing in advance would be a necessity. These are a few of many stars and celestial objects from which information could be transmitted in a reasonable timeframe. Since Jesus signifies the beginning of Pisces, and "(He is) with you until the end of the age (æon)," it is important that we dissect and discuss the origins of his light in a methodical manner.

Conceptual Age of Pisces, with the Heart of Taurus?

The ongoing debate about whether a soul enters the body at the moment of conception or at the first heartbeat has been a subject of deep discussion across scientific, religious, and spiritual communities. As we've examined the timing of Jesus' conception and the celestial alignment during that period, it would be remiss not to consider the possibility that a soul might enter the body at the first heartbeat. Given that science has yet to fully explain the initial cause of the heart's first beat, the notion that this moment could mark the entry of a soul into the body cannot be dismissed. However, this idea might seem unlikely when viewed through the lens of religious texts, particularly within Catholic and Christian traditions, as well as other religions, which often emphasize the significance of conception. But let's set aside these doctrinal perspectives for a moment and consider the scenario where the soul enters the body at the first heartbeat.

If Jesus was conceived on March 23rd, 30 BC, his first heartbeat would have occurred approximately five to six weeks later, placing it between

April 27th and May 4th, 30 BC. Notably, on May 1st, the Sun traverses the Taurus constellation, a region of the sky rich with significant star systems that pass behind the Sun during this time. The figure below illustrates the stars and cosmic objects aligned with the Sun around this date, offering a visual representation of the celestial events that could have coincided with this pivotal moment. The Sun moves from right to left in this figure as the year moves forward.

Gravitational Lensing and Cosmic Information Transfer

Gravitational lensing, a concept rooted in Einstein's General Theory of Relativity, occurs when light from a distant object is bent around a massive object, such as the Sun, due to the curvature of spacetime. This bending effect acts like a magnifying glass, focusing and intensifying the light—and the information it carries—from distant stars. This lensing effect could have focused and amplified the light from specific stars within the Pisces constellation, directing it toward Earth during the time of Jesus' conception, and coincidentally his resurrection.

Light carries information encoded in its frequency, amplitude, and phase. When light from distant stars is lensed by the Sun, this information is concentrated and directed toward Earth. During the period of Jesus' conception, it is possible that the stars in Pisces transmitted encoded information that influenced the development of the embryo, leading to what was a divine intervention or immaculate conception. God must have a way of communicating, influencing, and intervening with humanity throughout our existence. This is one method in which God would utilize the universe and all that exists within it to provide a version of God within a human body and soul in the most existential way.

In this context, Mary the mother of Jesus, could be seen as a conduit for divine and immaculate conception, and is understood and worshiped—albeit rightfully so—as a conduit for the son of God. The information carried by the light from these stars may have interacted with her at a cellular or even quantum level, imprinting upon the developing embryo the qualities that would define Jesus' extraordinary abilities and divine nature.

The Influence of Celestial Bodies on Human Development

Throughout history, the influence of celestial bodies on human development has been a topic of both scientific inquiry and mystical speculation. The alignment of planets, stars, and other cosmic objects has been thought to affect everything from personality traits to physical health. Stars emit electromagnetic radiation across a broad spectrum, including visible light, ultraviolet (UV) light, and infrared (IR) radiation. These frequencies can have various effects on biological systems, influencing everything from DNA repair mechanisms to hormonal regulation. The frequencies emitted by stars in the Pisces constellation, magnified and focused by the Sun's gravitational lensing, could have interacted with Mary's biological processes, affecting the genetic and epigenetic factors involved in Jesus' development.

Recent advances in quantum physics suggest that particles can become entangled across vast distances, meaning that a change in one particle can instantaneously affect another, regardless of the distance separating them. If one considers that the light particles (photons) from the stars in Pisces were entangled with particles within Mary's body, this could provide a mechanism for the instantaneous transfer of information, contributing to the divine qualities attributed to Jesus. This possibility cannot be ignored. This aspect of the theory is only applicable if the need for instantaneous communication is warranted. It is possible that a direct and instantaneous line of communication is not needed, since God and anything of a divine nature would not be binded by the limitations of time. Therefore, sending a signal many years prior to reaching the targeted destination would only require an understanding of cyclical alignments based on our timetable that both gravity and the warping of spacetime provide. DNA, and the design of the human genome by God, may have key aspects that can be utilized by immediate communication similar to quantum entanglement, to perform divine duties similar to immaculate conceptions. As we delve into the realm of the infancy of quantum computation and mechanics, and equally as time persists, progression in this field will most likely express this aspect in more detail if proven through science and technology. Until then, this aspect of science playing a crucial role in the subject must be acknowledged and acknowledged as only a potential aspect of my theory.

Cosmic Intervention and the Nature of Divinity

The idea that Jesus' conception was influenced by cosmic events and defined scientific understandings doesn't necessarily challenge traditional religious narratives, it offers a new perspective on the nature of divinity and how God provides assistance with his creations. Rather than seeing divine intervention as a purely supernatural and mysterious event, we can begin to calculate how the cosmos played an active role in shaping Jesus' identity, and plays a role in divine intervention overall. The alignment of stars within Pisces and their gravitational lensing by the Sun could be seen as God's way of transmitting divine energy and information to Earth and all those worthy that reside within it. In this view, the cosmos itself is in many ways a divine entity or conduit, capable of influencing the course of human history through its intricate web of physical and quantum interactions.

If Jesus' conception was indeed influenced by the light and information from distant stars, it suggests that he was not only a spiritual teacher and the true son of God—through his mindsets and teachings of Type II civilization and above—but also a cosmic being, connected to the universe in ways that transcend ordinary human experience. This perspective aligns with the concept of divine individuals utilizing the Universe around us to embody the consciousness of both God and the cosmos he created.

The Divine Hand in Cosmic Events

As we delve deeper into the relationship between cosmic events and the life of Jesus Christ, it is important to consider the perspective of many of the world's most intelligent physicists and coveted scientists. After dedicating their lives to understanding the universe through mathematics and scientific inquiry, many have come to acknowledge the existence of a higher power—God—who plays a role in the design and order of the cosmos. Throughout history, some of the most brilliant minds, including Albert Einstein, Isaac Newton, and more contemporary figures, have come to the conclusion that the intricate patterns and laws governing the universe are not merely coincidental but suggest the presence of a divine intelligence. Einstein himself famously said, *"God does not play dice with the universe,"* reflecting his belief in a structured, purposeful creation. These scientists see the fingerprints of God in the mathematical precision of the universe, from the laws of physics to the intricate patterns found in nature. The consistency of these patterns across vast scales—from the atomic to the cosmic—implies a design far beyond apparent human understanding, suggesting that the universe operates under a set of divine principles. However, we are here to dissect and discuss how these divine principles are expressed and implemented by divinity, or in a better term, by God. If we consider this mindset, it becomes plausible to believe that God, through the mechanisms of precession and the alignment of distant stars, orchestrates events that influence the course of human history. The precession of the equinoxes, which is a slow wobble in Earth's axis of rotation in relation to the stars and Universe surrounding it, gradually changes the orientation of the Earth relative to the stars creating cycles of time that have been associated with the rise and fall of civilizations.

The specific orientation of stars and the data encoded in the light they emit—whether through gravitational lensing, electromagnetic waves, or quantum interactions—could be the means by which God choreographs certain individuals to have an influential impact on humanity. Time on Earth is limited to those that reside within the planet, but not to those and anything that resides elsewhere—whether that be on Pisces or in a previous constellation that governs humanity's existentialism. This influence might be more pronounced at specific times when celestial alignments and cosmic energies converge in ways that are conducive to the emergence of divinely inspired figures like Jesus Christ. Considering the idea that God is intricately involved in the workings of the cosmos, it stands to reason that the conception and life of Jesus Christ were influenced by this divine order. The alignment of stars within the Pisces constellation during the time of Jesus' conception, coupled with the gravitational lensing effect of the Sun, may have been part of a deliberate act of divine intervention. Which was repeated for many millennia prior to his existence. However, the Pisces age, according to actual observance of the constellation, will not cease to exist for hundreds of years. Don't get your hopes up for an upcoming transfer of control through divinity by the cosmos or God. Science, mathematics, observed facts, and physics will confirm this statement.

Through the precise timing of celestial events, God could have ensured that the light and electromagnetic information from these stars reached Earth in a way that would influence Mary, the mother of Jesus, and the developing embryo. This cosmic influence could have imbued Jesus with the qualities that made him a figure of species-altering spiritual significance, destined to change the course of human history. Which means, regardless of our current "bath" of light, frequencies, and resulting vibrations in the modern world of computation and communication, the human species will never be able to combat divine beings that do not reside within our localized (planetary) timeline. We are bound to the planet that controls our clocks, and even if humanity controlled time itself here on Earth, it would not compete with the everlasting vastness and constant change in both space and time that the universe as a whole depicts to us when we deduce it in mathematics, and in observance of time. Humanity has a 200,000 year long history, and our planet is over 4.5 billion years old. Our planet is rather young in comparison to the cosmos and God. We will never "leap" ahead of either, and toying with either is a losing battle that both can see well before our attempts.

The Role of Surrounding Individuals in Jesus' Life

While the divine imprinting upon Mary ensured that Jesus was born with the specific qualities and potential necessary to fulfill his role, the broader context of his life required additional divine intervention. Humanity itself needed to recognize and nurture these qualities, and to guide Jesus on the path that would lead him to become the spiritual teacher and savior as foretold. It was not enough for Jesus to be born with inherent divine qualities; the people around him also played a crucial role in shaping his journey. Divine intervention extended to the lives of those who would influence Jesus, ensuring that he encountered the right individuals at the right times to learn the lessons he needed. From personal experience and utilizing common sense, the individuals that influenced, shaped, and thought Jesus were not simply relying on themselves. They were guided by the divine light and frequencies of those that exist beyond humanity's grasp. From his family, such as Mary and Joseph, to the teachers, scholars, and spiritual leaders he would encounter, each person contributed to Jesus' development. Their influence, guided by divine intervention, helped Jesus to understand

and cultivate the gifts and knowledge imprinted upon him, enabling him to fulfill his destiny.

According to various historical and esoteric accounts, Jesus traveled extensively during his early years, visiting places such as Egypt, Asia Minor (Turkey), India, and possibly Tibet. These travels were not merely for exploration but were divinely orchestrated journeys designed to expose him to a wealth of spiritual wisdom and knowledge. In these distant lands, Jesus encountered different philosophies, practices, and teachings that would later form the foundation of his own message. Whether learning from the mystics of India, the soon-to-be churches in Asia Minor or Turkey, the scholars of Alexandria, or the monks of the Far East, these experiences allowed Jesus to integrate a wide range of spiritual knowledge and overall wisdom of humanity as a whole, which he then synthesized into a cohesive and transformative teaching. This knowledge, combined with the divine qualities inherent in him from birth, allowed Jesus to speak with authority on matters of the spirit, healing, and the nature of God. His teachings resonated with people across cultural and religious boundaries, demonstrating the universal nature of his message.

The Influence of Human Nature on Jesus' Life and Death

Despite the divine intervention and the specific wisdom that Jesus possessed, he ultimately faced the harsh realities of human nature. Throughout history, humanity has often reacted with fear, hostility, and violence to those who challenge established norms or present a message that threatens the status quo. In Jesus' case, the wisdom and truth he shared were met with both acceptance and rejection. While many followed him, recognizing the divine truth and potential for human progression in his teachings, others—particularly those in positions of power that had succumbed to the indoctrination of humanity—saw him as a threat. The very qualities that made Jesus a beacon of spiritual light also made him a target for those who feared change. In the end, it was the darker aspects of human nature—ignorance, fear, and malice—that led to his crucifixion. Despite the divine guidance that had shaped his life, Jesus' fate was sealed by the same human flaws that have led to the persecution of many enlightened figures throughout history.

The Crucifixion, Forgiveness, and the Defeat of Evil

The crucifixion of Jesus Christ is one of the most significant and profound events in human history. It represents not only the physical suffering and sacrifice of a divine being but also a spiritual battle against the forces of evil that have plagued humanity since the dawn of time. Through his crucifixion, Jesus demonstrated the power of forgiveness, the strength of divine love, and the ultimate defeat of evil. As Jesus hung on the cross, suffering unimaginable pain and betrayal, he chose to forgive those who had wronged him. This act of forgiveness was not merely a gesture of kindness; it was a powerful spiritual act that warded off the evils that sought to consume him and, by extension, all of humanity. In many religious traditions, there exists a figure known as Satan or the Devil, who is believed to be a prisoner and the one inevitable warden of hell. This entity's goal is to ensnare human souls by leading them into sin, as defined not only by the Christian and Catholic Churches but by numerous other religious and spiritual systems. The Devil seeks to corrupt humanity, leading souls to eternal damnation in Hades. Jesus, with his divine spirituality, intellectual mindset, and deep understanding of humanity as a whole, knew what needed to be done upon his death. His forgiveness on the cross was a direct affront to the Devil's plans, a refusal to succumb to the hatred, anger, and despair that the forces of evil sought to instill in him.

Combined with the help of divine light and frequencies, he was never fighting a losing battle when carrying his cross up the hill and being nailed to it.

The Forty Days of Spiritual Crucifixion and Trials

The period during which Jesus was spiritually crucified is often symbolically referred to as forty days and forty nights. Similar to the forty days and nights spent in the desert, this time was also marked by a series of spiritual trials, as various individuals from society visited him—not to offer comfort or solace, but to mock, taunt, and attempt to break his spirit. These individuals were not empathetic in the true sense, but they possessed a certain sensitivity to the emotions and energy of others. They fed off the energy that Jesus exuded, both from the physical torment of being physically crucified on the cross, and the emotional and spiritual pain of enduring false judgments and betrayals. These individuals, driven by their own darkness, acted like Anubis in ancient Egyptian mythology, testing the weight of Jesus' heart. However, it wasn't to aid Jesus into the afterlife properly, it was to feast on him in ways that have been practiced for millennia. Despite their attempts to sway him into committing to the sins recognized by various religions, Jesus remained steadfast in his forgiveness and divine purpose. Not only was the will of his spirit which was developed throughout his life unbreakable, but the help from divine powers and the almighty God, his father, assisted him in his darkest and most challenging days. Regardless of religious and spiritual beliefs, the strength that Jesus expressed during these days cannot be ignored. Jesus was one of, if not the only, individual that was able to withstand the spiritual trials and spiritual crucifixion for more than a few weeks, and survive being nailed to a cross after being whipped, all while surviving over six hours or so. I want the reader to perceive being nailed by both hands and feet to a cross after constant abuse for almost forty days and nights. With no food, no water, and no comfort provided during the physical crucifixion. Individuals within his society would come to break him, feed off him, and damage his body before and while being nailed to the cross. Most people that live in the modern world would cede in an attempt to be let free. He did not. The strength Jesus displayed in these moments was not merely physical but deeply spiritual. He understood that to give in to despair, anger, or hatred would be to surrender to the very forces he was divinely designed to, and ultimately arrived to defeat. He would surrender to evil at the expense of his teachings to humanity and overall spiritual or religious beliefs. By holding fast to his faith and forgiving those who sought to harm him, Jesus ensured that evil would not claim victory over him or the world that existed after him.

As these individuals approached Jesus to throw stones and hurl insults, they were not just expressing their disdain; they were feeding off the energy exuded off a man who was being ostracized and hunted by the Pagans and Hebrews or Jews in power. This occurred both during the spiritual crucifixion and trials, and the physical one *seemingly* at his weakest. I have personally experienced this behavior, even in today's world. In today's terms, this behavior might be compared to the highs experienced by those who engage in cruel or malicious acts—a temporary surge of power at the expense of another's suffering. Many individuals would be hard-pressed to maintain their anger and aggression without clenching their fists or pressing their fingers in certain ways while experiencing a "rise" and sense of euphoria. My suggestion is to tell anyone who is emotionally arguing or yelling, to simply open both hands wide and extend their fingers outward, maintaining separation of all of the fingers. If they can maintain an open hand without gravitating any of their fingers together, they will most likely calm down. But if they inadvertently clench without attempting to refrain from clenching, you know what you are in front of. It's a tricky test, since the feeling of euphoria and adrenaline is difficult for weak minded individuals to overcome. The peaceful can do this at ease.

Despite society's attempts to sway Jesus into committing to the sins of hatred, he remained legendary in his forgiveness and divine purpose. Expanding upon this concept, individuals in today's world who attempt to "cook" or feed off another person's dismay, depression, and sadness can do so not only through technological means but through the very medium that surrounds us all—our own biofield and electromagnetic energy. By manipulating their own biofield, such individuals can forcefully tap into a section of their heart and emotions, using hand gestures as a conduit to extend and intensify the feelings of despair, depression, and sadness in others they are connected to. This phenomenon is nothing new and is why certain hand gestures are commonly seen in illustrations and iconography throughout history. Satan and the Devil are often depicted with the index finger pressing against the thumb, symbolizing the act of channeling and prolonging negative emotions for personal gain. It is not just the devil, but various religious and spiritual beliefs systems that utilize this methodology. Conversely, Jesus is often depicted with the ring finger pressing against the thumb, representing the act of shortening or eliminating these malevolent influences.

Figure 5. This is a typical hand gesture performed by Jesus, it depicts his name in Greek, ICXC, but also means blessing, in the presence of

Jesus Christ, and can both cancel out the usage of the thumb to connect to the index finger—channeling the part of your heart that does not reside in negative mindsets—while warding off attempts of forceful other gestures. It is capable of overpowering all other hand gestures when the faith and confidence exists in the person.

Figure 6. The above two photos depict the hand gesture for 666 or 616 (original number of the beast from Papyrus 115). It signifies the mark of the beast in some beliefs, or invoking the evil eye in others. It is also used by white supremacists. The index finger to thumb gestures were utilized during Jesus' crucifixion, individuals that attempted to feed off his misery would squeeze their index fingers into their thumbs in various gestures, trying to prolong the feeling exuded by Jesus Christ upon death. Regardless, both channel a section of your heart and chakras that bring negative thoughts and feelings that can be prolonged when applied to others you are connected to or within their biofield. Some religions do not acknowledge evil in the same way, and use it for sparking enough energy and emotion to traverse the medium during meditation and intense contemplation. In terms of invoking the evil eye upon others, it isn't as simple as making the gesture. Those that have lived a more honest and kind life will not be affected in the long or short term by individuals that try to invoke the evil eye upon them. The opposite actually occurs, even if others try to force the gesture or other actions upon those that are kind and protected. But it does expose the evils that are being portrayed.

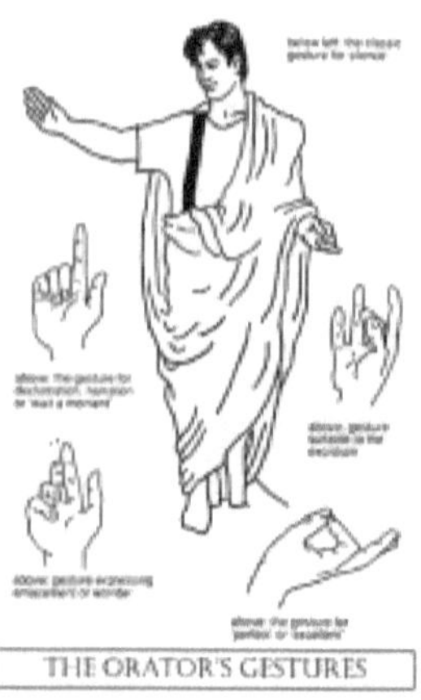

Figure 7. Here are many gestures performed by various members of the Catholic Church. Notice how in the bottom right of the image they took the 666 / 616 gesture and flipped the hand so that they can utilize the feeling it provides. The bent wrist and squared angle of the index finger changes the impact of external forces, signifying perfection in taking advantage of the evil eye or the evil individuals that traverse the medium around us.

The difference between these gestures lies in their impact on the heart, chakras, and the biofield. The index finger to the thumb gesture, when the wrist is not bent and the index finger is not sharply bent, is used to channel into a section of the heart and biofield to extend and intensify negative emotions and energy for selfish purposes. In contrast, the ring finger to the thumb gesture is intended to counteract this effect, reducing or eliminating the negative influence and restoring balance to the heart and biofield.

This concept provides insight into the realms of heart and brain coherence, emotional intelligence, and the broader function of the human heart as an electromagnetic center. It highlights the importance of the heart in regulating not just emotional well-being but also the overall energy and health of the human body. While these actions may not be purely spiritual, they are deeply intertwined with the electromagnetic field of the human body, influencing the way energy is transmitted and received within ourselves and in our interactions with others. It is also a way to utilize each and every human being to abide to the specific instructions of malicious individuals. These are not deities or divine individuals, but rather humans and evil entities that control them. Because it is fueled by sinful manners, mindsets, and aspirations. It is not fueled by those that comprehend the positive and everlasting progression of humanity and human nature. Destructive mindsets lead to destruction. Jesus, through his deep understanding of these dynamics, maintained a powerful heart coherence that prevented the negative energies directed at him from taking hold. His ability to forgive and remain aligned with divine love ensured that his heart remained pure and unburdened, even in the face of such intense spiritual and emotional trials. What he learned over the 18 year gap in historical texts when visiting countries like Egypt, Asia Minor, and India, showed him how to overcome the mark of the beast and all those which utilize such energies.

The Three Days in the Tomb and the Harrowing of Hell

After Jesus' death, there was a three-day gap before his resurrection. These three days were a period of complex spiritual significance and divine planning. During this time, Jesus descended into hell—a place traditionally governed by Satan and his demons, where the souls of the damned were believed to be imprisoned. In Christian theology, this event is known as the "Harrowing of Hell." It was during these three days that Jesus confronted the warden of hell with the attributes instilled by the divine intervention of God and the teachings he ensured Jesus experienced. Jesus was not alone in this dark place: the light of God was with him, providing the strength and divine authority needed to overcome the forces of evil. As Jesus traversed the depths of Hades he encountered souls trapped there; souls that were deserving of his knowledge and salvation he possessed, but they lacked the divine æon-specific intervention, and spiritual journey that had shaped his life. Jesus, in his infinite and reasonable compassion, offered these souls redemption, bringing back with him those who were trapped alongside the damned. Although the amount of souls saved from eternal damnation were limited, there is a reason why this is so. Jesus has forgiving nature, but if a soul has been damned to Hades the only opportunity for salvation was being wrongly sentenced to Hades, or the sheer fact that previous religious leaders had one place to reside upon death during their age: Sheol. One example of such influential figures was Moses. We all know his story of perseverance and divinity.

While Jesus' Harrowing of Hell and his resurrection brought back a minimal number of souls from the previous age, the significance of this act was far-reaching. These souls were not just the beneficiaries of his divine compassion but were also symbols of the hope and salvation that Jesus offered to all of humanity. More importantly, Jesus' actions laid down the foundation for the souls of those who lived after him. Through his teachings, his example of forgiveness, and his victory over death, Jesus ensured that the deceit and entrapments of evil could no longer claim the souls of the people who followed his path. The beautiful souls that lived after him were thus given the tools and the spiritual strength to resist the temptations and evils that had plagued previous generations. Jesus' influence established a legacy that allowed his followers to maintain purity of heart and spirit. By embracing his teachings and living in accordance with his example, these souls could navigate the challenges of life in a way that aligned them with the divine, ensuring that their hearts remained light and unburdened by the sins and evils of the world.

This foundation was critical for humanity as it moved further into the Age of Pisces. As the age progressed, the values that Jesus imparted—compassion, positivity, forgiveness, love—became cornerstones of spiritual life, guiding countless individuals toward enlightenment and salvation. His teachings served as a spiritual compass, helping humanity to steer clear of the darkness and toward the light.

The Keys to Death and Hades

Upon his resurrection, Jesus emerged not only as the savior of humanity but as the one who holds the keys to death and Hades. This perspicacious symbolism represents his victory over the forces of evil and his authority over life and death. The Devil, who once sought to claim the souls of the damned, was defeated by Jesus' unwavering faith and guidance of the light, forgiveness, and divine nature. Through his death and resurrection, Jesus shattered the powers of Hades ensuring that the keys to life, death, and the afterlife were in the hands of the one who embodied divine love and compassion, rather than deceit and governance of human emotions. It was no longer in the hands of those seeking to feed off of any soul that can exude emotion in a way that can be feasted on.

The Age of Pisces, Jesus' Eternal Influence, and Overcoming the Aggression of the Previous Age

The significance of Jesus Christ's life, death, and resurrection extends far beyond the immediate context of his time. His existence is deeply intertwined with the cosmic cycles, particularly the Age of Pisces. This age, marked by the symbolism of the fish, is often associated with spiritual awakening and compassion. Jesus was conceived and born at the very beginning of the Age of Pisces, a time of significant spiritual intelligence. The transition from the previous age, often associated with the Aries constellation and symbolized by the ram, represents a shift from a period of war and conquest to an era focused on spiritual enlightenment and the inner life. Jesus' influence and teachings will endure until the end of this age, or between 2150 and 2605 AD. The entities that held sway over the afterlife during this time were often seen as harsh, demanding, and unforgiving. The souls trapped by the evils of the previous age were often subjected to the deceit and cruelty of these forces. Jesus, by contrast, came into the world at the dawn of a new age, one in which the potential for spiritual redemption and compassion was heightened. Through his life and ultimate sacrifice, Jesus overcame the evils that had dominated the previous age, superseding all entities that controlled the darker aspects of the afterlife. His victory over death and Hades was not just a personal triumph but a cosmic one, signaling the beginning of a new spiritual era.

Jesus' promise to be with us until the end of the age is not just a spiritual assurance but a reflection of the enduring power of his teachings. As the Age of Pisces continues, his influence remains a guiding light for billions of people around the world. The principles of compassion, forgiveness, and love that he championed continue to shape human consciousness, leading people toward a deeper understanding of their own spiritual nature. These principles are also necessities for the proper development and progression of humanity, for without them intelligent life forms of any sentience will inevitably experience total destruction and demise. The transition to the next age of Aquarius will bring new challenges and opportunities for spiritual growth. However, the foundation that Jesus laid during the Age of Pisces will continue to serve as a cornerstone for those who seek to live in alignment with divine principles, and a society which progresses safely.

Ancient Wisdom, Cosmic Intervention, Anubis, and the Weighing of the Heart

The spiritual teachings of Jesus Christ, while often viewed through the lens of Christian doctrine, share deep connections with ancient wisdom from various cultures. One of the most striking parallels can be found in the beliefs of ancient Egypt, where the afterlife was governed by the Gods, and the heart was the key to one's eternal fate. Understanding this connection sheds light on the universal nature of Jesus' message and his triumph over evil.

In ancient Egyptian culture, the god Anubis played a crucial role in the afterlife. He was tasked with weighing the hearts of the deceased against the feather of Ma'at, the goddess of truth and justice. If a heart was heavier than the feather, burdened by sin and wrongdoing, it would be devoured by Ammit, the soul-eater, condemning the individual to oblivion. But if the heart was lighter than the feather, it signified a life lived in harmony with Ma'at, and the soul would ascend to the heavens. This concept of the heart's weight determining one's fate after death is echoed in the teachings of Jesus. His messages of love, forgiveness, and compassion were not just moral guidelines but spiritual tools designed to lighten the heart. By instructing his followers to "love thy neighbor" and "forgive thy neighbor," Jesus was offering a path to spiritual liberation—a way to ensure that their hearts would not be weighed down by anger, hatred, or unforgiveness. However, Jesus' teachings were clear in distinguishing between forgiving people and condoning evil. His message was never about loving or accepting evil, but about overcoming it by maintaining purity of heart even in the face of adversity. This was a crucial difference from previous teachings and a key element in the cosmic intervention that shaped Jesus' life.

The divine intervention that marked Jesus' life was not just about ensuring his spiritual purity but about preparing him to confront the forces of evil that had long held sway over humanity. These forces, often associated with specific frequencies, resonances, emotions, and waveforms could influence individuals, leading them down paths of darkness and despair. Throughout his life, Jesus learned to recognize and combat these negative influences. His teachings were more than ethical instructions; they were spiritual defenses against the resonance of evil. By cultivating positive vibrations and staying aligned with divinely instructed and choreographed frequencies, Jesus showed humanity how to maintain their spiritual integrity and resist the pull of darkness. Yet, maintaining a "positive light" or spiritual purity was not enough on its own to ensure salvation. The lightening of one's heart—a concept central to both Egyptian beliefs and Jesus' teachings—required more. It required the kind of relentless spiritual resilience and unwavering faith that Jesus demonstrated during his crucifixion.

The Victory Over Anubis and the Triumph of the Son of God

In overcoming the trials of the cross, Jesus not only fulfilled his mission but also symbolically defeated the forces represented by Anubis. Where Anubis weighed the hearts of the dead, Jesus had his heart weighed by the living—by those who sought to break him. And in the end, it was Jesus who emerged victorious, his heart unburdened and his soul intact. This victory was more than just a personal triumph; it was a divinely intervened and instructed one. By maintaining the lightness of his heart, Jesus ensured that he could not be claimed by the forces of evil, whether they resided in this world or the next. His ability to forgive, even those who sought to harm him, demonstrated the ultimate power of divine love—a power that transcends all earthly and spiritual realms. This is why Jesus holds the keys to death and Hades, a position of authority that reflects his victory over the forces of darkness. His life, death, and resurrection are a testament to the idea that love, compassion, and forgiveness are the true paths to spiritual freedom and eternal life. By following his example, humanity can ensure that their hearts remain light, free from the burdens of sin and evil, and aligned with the divine will.

The teachings of Jesus Christ are deeply rooted in both ancient wisdom and cosmic alignments. His message of love and forgiveness not only aligns with the spiritual practices of previous ages, such as those of ancient Egypt, but also represents an evolution in humanity's understanding of divine will. By overcoming the trials of the cross and maintaining the lightness of his heart, Jesus demonstrated the power of divine intervention and the importance of spiritual resilience. In recognizing the interconnectedness of his teachings with those of previous ages, we gain a deeper appreciation for the universality of Jesus' message. His life serves as a bridge between the ancient and the modern, between the physical and the spiritual, offering a path to salvation that transcends time, culture, and religious boundaries. Through his example, Jesus not only secured his rightful and appointed place as the Son of God, but also provided humanity with the means to overcome the evils that have long plagued our world. As humanity moves through the ages, his teachings will continue to guide us, ensuring that our hearts remain light, our spirits strong, and our connection to the divine unbroken. All while providing a foundation to what is necessary for humanity to proceed with their evolution and progression into the next stages of intellectual civilization.

Notes

Chapter 4: Oh Pharaoh Tesla Stop Eavesdropping...

...But Write It All Down So Trump Can Build Hotels With Your Discoveries...

The relationship between Nikola Tesla's work and the ancient pyramids is one that has intrigued researchers and enthusiasts alike. Both represent monumental achievements in their respective eras, and both share an enigmatic connection to energy, transmission, and the harnessing of natural forces. Tesla, who was fascinated by the pyramids, often alluded to their potential as giant energy transmitters or receivers, and his own work with Wardenclyffe Tower at Colorado Springs seems to echo some of these ancient principles. Tesla's Wardenclyffe Tower was built with the intention of transmitting wireless energy across great distances. Its design, however, wasn't just about functionality; it bore a striking resemblance to ancient structures, particularly most pyramids, with its integration of the Earth's natural energy fields. Many important monolithic pyramids and Tesla's tower were constructed over natural aquifers, utilizing the Earth's magnetic and electromagnetic fields. Tesla believed that these natural elements could amplify the transmission and reception of energy, an idea that finds an uncanny parallel—specifically in the construction of the Great Pyramid of Giza.

The Much More Important and Intellectual 33rd Degree of Masonry

The Great Pyramid of Giza, perhaps the most famous of these structures, sits on the 33rd-degree parallel—a latitude that has long been associated with mystical and energetic significance. This latitude, in particular, is known for electromagnetic anomalies, which may not be coincidental. The Great Pyramid is also located on one of the world's most significant ley lines, often referred to as the "world grid." This ley line is believed to connect several other ancient and sacred sites, including Stonehenge and the statues of Easter Island. But the Great Pyramid's location is not its only notable feature. Beneath it lies a significant aquifer, a natural underground water source that plays a crucial role in the pyramid's potential to generate and amplify natural electromagnetic energy. The combination of the pyramid's precise alignment with the cardinal points and its strategic placement over this aquifer might have been deliberately chosen to enhance its energetic properties, enabling it to interact with the Earth's natural energy grid. Electromagnetic anomalies and other unusual phenomena have been reported at several pyramid sites around the world, adding another layer to the proof of an overall truthful intelligence and profound understanding of—both the cosmos and the Earth itself—by ancient civilizations. These anomalies suggest that the locations of the pyramids were not chosen arbitrarily, but rather were carefully selected for their unique energetic properties and environmental phenomena.

The Pyramids of Giza

The Great Pyramid of Giza, long thought to be merely a tomb for pharaohs, has been reconsidered by some as a possible ancient power plant. This theory suggests that the pyramids, particularly the Great Pyramid, were designed to harness and distribute energy on a massive scale. The pyramids are strategically placed on the Earth's grid, a network of energy lines known as ley lines. This positioning, combined with their construction materials, such as limestone and granite rich in quartz, suggests that they could have been used to channel and amplify natural energy fields.

At the apex of the pyramids, it is believed there once resided massive quartz crystals, possibly encased in copper or gold, which would have been capable of focusing and directing energy. The inner chambers of the pyramids, such as the King's Chamber, are believed to have been resonating chambers, designed to generate and amplify vibrations. The alignment of the pyramid with specific star systems, particularly Sirius, adds another layer of intrigue, suggesting that these structures were also intended to connect with celestial energies. The ancient pyramids, strategically positioned around the globe, form more than just architectural marvels; they are a part of a vast, interconnected network that takes full advantage of the medium in which everything resides. This medium, an invisible yet omnipresent field that binds all existence, allows for the transmission of energy, information, and communication not only across the Earth but also with the cosmos beyond our atmosphere. Researchers have documented various electromagnetic anomalies at the Great Pyramids of Giza. One of the most intriguing findings is that the pyramid seems to focus electromagnetic energy, particularly in the microwave frequency range, within its chambers and beneath its base. This effect could be linked to the pyramid's precise geometrical structure, which might interact with the Earth's electromagnetic field in a unique way. Additionally, the area around Giza is known for its strong geomagnetic forces, which could be related to the pyramid's alignment with the cardinal points and its location on a significant ley line.

Teotihuacan, Pyramid of the Sun

Thousands of miles away from Egypt, in what is now Mexico, lies the Pyramid of the Sun at Teotihuacan. This pyramid, too, is aligned with ley lines, particularly in relation to other ancient Mesoamerican sites. The builders of Teotihuacan were likely aware of underground water systems, as recent discoveries have revealed subterranean tunnels and chambers beneath the Pyramid of the Sun that may have been connected to water sources. These features likely played a role in the site's ritualistic and energetic significance, linking it to the broader network of energy that crisscrosses the globe. The Pyramid of the Sun in Teotihuacan has also been associated with electromagnetic anomalies. Studies have shown that the pyramid, which is aligned with certain astronomical features, might amplify natural energy fields in the area. Some researchers have speculated that the extensive network of underground tunnels and chambers beneath Teotihuacan could enhance these effects by channeling and focusing electromagnetic energy. Visitors to the site have reported experiencing sensations such as dizziness and heightened awareness, which could be related to these energetic phenomena.

The Qinling Pyramids, China

In China, the Qinling Pyramids near Xi'an are thought to be positioned along ancient Chinese meridians—conceptually similar to ley lines in their connection to energy flow and alignment with the stars. While less is known about aquifers beneath these pyramids, the principles of feng shui, which emphasize the importance of water in harnessing and balancing energy, suggest that these pyramids, too, were built with an awareness of the Earth's natural energies. The Qinling Pyramids in China are located in an area with significant geomagnetic anomalies. The region is known for its complex geology, including fault lines and mineral deposits that could influence the local magnetic field. These factors might have contributed to the selection of the site for pyramid construction. Some studies suggest that the pyramids' orientation might be designed to harness these natural energies, enhancing the spiritual or ritualistic purposes of the structures.

Cholula Pyramid, Mexico

The Great Pyramid of Cholula, the largest by volume in the world, also fits into this global network. Aligned with ley lines connecting other significant Mesoamerican sites, its placement and construction were likely designed to harness the spiritual and energetic forces of the Earth. Though specific aquifers beneath Cholula have not been widely documented, the significance of water in Mesoamerican rituals strongly indicates that such considerations were at play. At the Great Pyramid of Cholula, the electromagnetic anomalies are less well-documented but no less intriguing. The pyramid's massive size and its location within a highly volcanic region could influence local electromagnetic fields. Volcanic activity is often associated with the generation of strong electrical fields, and the pyramid's alignment with other Mesoamerican sites suggests that its builders were aware of these forces. The pyramid's location on a ley line could further amplify these effects, creating a powerful energetic hub.

The Bosnian Pyramids

Even the controversial Bosnian pyramids, if indeed they are man-made structures, are believed by some to be aligned with ley lines that connect other significant sites. These pyramids are also said to be situated above natural aquifers, which proponents argue could amplify the energy and healing properties attributed to the site. The Bosnian pyramids, although controversial, are claimed to exhibit significant electromagnetic anomalies. Proponents argue that these pyramids generate intense energy fields, which are detectable even with basic measuring equipment. Some researchers have reported finding a focused beam of electromagnetic energy rising from the top of the largest pyramid, the so-called Pyramid of the Sun. This energy is said to be much stronger than what would be expected from natural geological formations, leading some to believe that the pyramid was intentionally designed to harness and amplify these energies.

The Hathor Hieroglyphs and the Baghdad Batteries

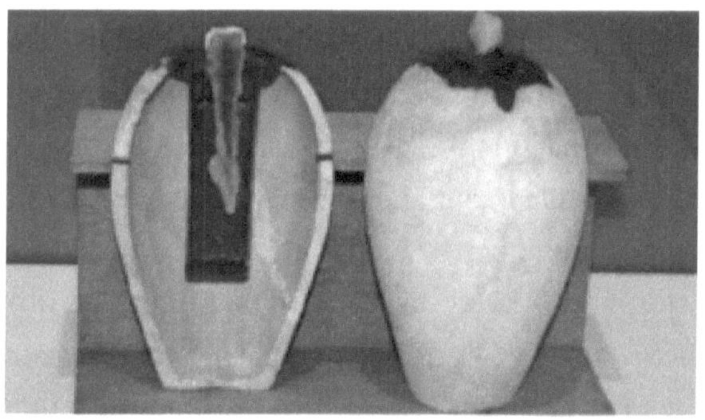

The ancient Egyptians left behind a wealth of knowledge encoded in their hieroglyphs, and among the most intriguing are the depictions found in the Temple of Hathor at Dendera. These hieroglyphs appear to show large bulbs connected to what some believe to be early forms of batteries, often referred to as Baghdad batteries. The Baghdad battery is an ancient artifact discovered in Iraq, consisting of a clay jar with a copper cylinder and an iron rod, believed to have been capable of generating a small electric charge when filled with an acidic liquid. These depictions suggest that the ancient Egyptians might have understood and harnessed electricity long before it was officially 'discovered' in modern times. The hieroglyphs found in the Temple of Hathor, which was a significant center for the study of science, physics, and mystical arts, indicate that the pharaohs might have used this knowledge to portray godlike behavior, further cementing their divine status in the eyes of their subjects.

Ley Lines, Electromagnetic Anomalies, and Holes in the Electromagnetic Field

The presence of ley lines at these pyramid sites is thought to contribute to the electromagnetic anomalies observed. Ley lines are believed to be paths of concentrated energy that crisscross the Earth, connecting ancient and sacred sites. The intersection of ley lines at pyramid sites could create energetic hotspots, where electromagnetic fields are stronger or more focused. These intersections might also serve as conduits for the transmission of energy and information across vast distances. In some cases, pyramid sites are associated with anomalies in the Earth's electromagnetic field, such as holes or disruptions. These anomalies might be caused by the interaction between the pyramids' structures and the natural geomagnetic forces of the Earth. For example, the precise alignment and shape of the pyramids could create interference patterns in the electromagnetic field, leading to areas of increased or decreased energy. Such anomalies could have been used by ancient civilizations for various purposes, including spiritual rituals, healing practices, or even communication with the cosmos.

An Understanding of the Universe's Network of Communication

These pyramids are not isolated monuments but are part of a global network that taps into the Earth's natural energies. The alignment of these structures with ley lines and their connection to underground aquifers suggest that ancient civilizations were highly attuned to the medium—the visually invisible field of energy that surrounds and permeates everything. By placing these pyramids along ley lines, ancient builders may have been able to tap into and amplify the Earth's energy, creating focal points that could transmit information, energy, and perhaps even communication across vast distances. But the influence of these pyramids might extend beyond the Earth. The strategic placement of these structures and their interaction with the Earth's electromagnetic field could have allowed them to act as beacons, sending and receiving information not only within the planet but also out into the cosmos. This network, stretching across continents and cultures, may have been designed to connect humanity with the universe, utilizing the natural medium of energy and light to bridge the gap between the Earth and the stars. In this way, the pyramids can be seen as ancient communication devices, utilizing the natural properties of the Earth and the medium that surrounds us to send messages across space and time. They stand as a testament to the knowledge and understanding that ancient civilizations had of the natural world—an understanding that allowed them to create structures that continue to baffle and inspire modern society to this day.

Tesla's Wardenclyffe Tower

Tesla's Wardenclyffe Tower was an ambitious project intended to revolutionize the transmission of energy and communication. Built over an aquifer, the tower was designed to harness the Earth's natural energy, much like the pyramids. Tesla believed that by tapping into the Earth's resonance, he could transmit energy wirelessly across the globe, providing free power to everyone. The tower was constructed with a deep well beneath it, intended to create a powerful oscillating circuit with the Earth. Tesla's understanding of the Earth's electrical properties led him to believe that this natural resonance could be used to transmit energy without wires, a concept that echoes the ancient Egyptian use of natural materials and positioning to harness energy. Despite the potential of Wardenclyffe, Tesla's project was ultimately shut down, and the tower was dismantled. However, the principles behind it have continued to inspire modern research into wireless energy transmission and the possibility of harnessing natural energy fields.

The Role of Natural Aquifers and Electromagnetic Fields

Both Tesla's tower and the pyramids were built over natural aquifers, suggesting a deliberate attempt to harness the Earth's natural energy. Water is a highly conductive material, and the presence of aquifers beneath these structures could have played a crucial role in amplifying their energy capabilities. The interaction between the Earth's magnetic field, the conductive properties of water, and the resonating structures above could have created a powerful energy system. Tesla's understanding of the Earth's electrical potential led him to design systems that could tap into these natural energy sources. Similarly, the ancient Egyptians' use of materials rich in quartz and the precise alignment of the pyramids with celestial bodies suggests a sophisticated understanding of how to harness and amplify natural energy.

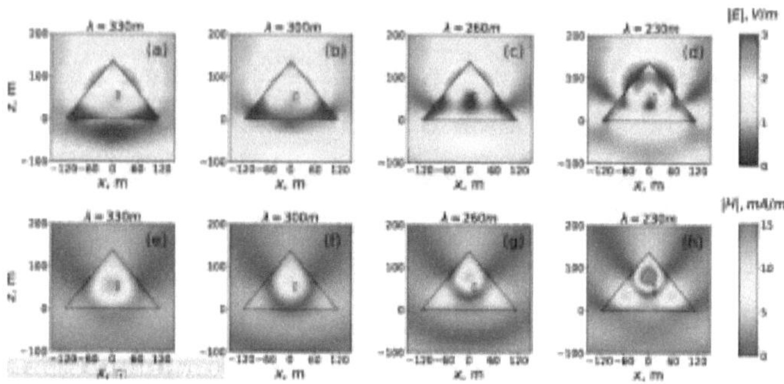

Figure 1. *Distributions of electric (top row) and magnetic (bottom row) field magnitudes in the free space are shown*

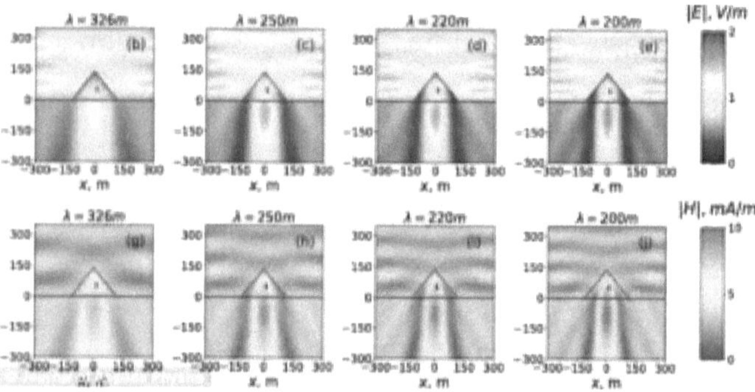

Figure 2. *The distributions of electric (a)–(e) and magnetic (f)–(j) field magnitude in the Pyramid and its supporting substrate is shown above*

They Even Had Tesla Live in a Hotel Until His Death

After Tesla's death, many of his papers and research notes were seized by the U.S. government, under the guise of national security concerns. One of the key figures involved in reviewing Tesla's work was John G. Trump, an MIT professor and the uncle of former President Donald Trump. John G. Trump claimed that Tesla's papers contained nothing of significant scientific value. However, the decision to keep Tesla's work hidden from the public has fueled ongoing speculation that Tesla's research may have contained revolutionary ideas that were not yet ready for the world to see—or that were considered too dangerous to release. Tesla was not only watched but also scrutinized throughout his life, and his eventual fate seemed to be one of isolation and containment, similar to the very energy he sought to understand and harness.

The Quartz Crystal Connection

Quartz crystals are known for their piezoelectric properties, meaning they can generate an electric charge when subjected to mechanical stress. This property makes them ideal for use in both ancient and modern technology. The use of quartz in the construction of the pyramids, particularly in the inner chambers, suggests that the ancient Egyptians understood these properties and used them to create a resonating energy system. Tesla, too, was fascinated by quartz and other crystals, incorporating them into his designs to enhance the transmission and reception of energy. The idea that the pyramids could have served as giant energy transmitters, using quartz crystals to focus and direct energy, aligns with Tesla's vision of wireless energy transmission.

Altermagnetism: The New Frontier

In recent years, researchers have discovered a new type of magnetism known as altermagnetism, which shows that certain materials, including crystals, can exhibit unique magnetic properties. This discovery has key implications for our understanding of energy transmission and storage, as well as the role that crystals could play in future technologies. Altermagnetism suggests that crystals like quartz could be used not only to generate and transmit energy but also to store information. This aligns with the idea that the pyramids, and perhaps Tesla's tower, could have been designed as not just energy transmitters but also as information storage systems. The potential for crystals to hold and transmit data opens up new possibilities for both ancient and modern technologies.

Orbs in the Great Pyramids and Throughout Egypt

The discovery of orbs within the Great Pyramid and at various sites across Egypt adds another layer of mystery to these ancient structures. These orbs, often considered to be dense pockets of information stored within light, may represent a form of ancient data storage, possibly linked to the electromagnetic fields and energy systems of the pyramids. Orbs have been a subject of spiritual exploration for many years, and their appearance in such a significant historical context suggests that they might be more than just anomalies. They could be seen as manifestations of the energy and information encoded within the pyramids, interacting with the unique electromagnetic environment of these ancient structures. This concept of orbs as carriers of information aligns with the idea that the pyramids were not just physical structures but also centers of knowledge and energy, designed to connect with both the Earth and the cosmos.

Now that we understand how light can be encoded with information, it is important to note that the Egyptians were rather advanced in masonry, electromagnetic abilities, and geometry. One theory is that orbs are spirits, or remnants of life itself. Thus, considering the pyramids were not tombs, but either magnetic or electromagnetic monoliths and structures, one can theorize how the geomagnetic and electromagnetic anomalies coupled with their structure and location could theoretically harness and store information over vast distances and time. Crystals hold charge and information, therefore, if the pyramids have ample crystal and/or a giant crystal at the top of each pyramid, they can potentially store information about a human being and the data they transfer throughout the medium. Granted, the stones used throughout the great Pyramids actually contain quartz—an essential and important crystal when speaking about information and energy storage. I will speak about neurons and neurites later on in the book, but essentially they are antennae the body and mind uses for memories and emotions stored within the medium. Neurons and neurites are proven to be necessary with memories, but don't actually store any type of information or data. This is something both the brain and heart have, and can be expressed by taking neurons or organs without data storage out of one body and placing them into another, effectively transferring memories to the recipient without the written data. When the heart is transplanted it does not bring any data storage with it. Placing data storage within DNA aside, it would be foolish to think data storage can't occur through the means of crystals, using something as powerful as the pyramids as a means to transmit and potentially store said data. Orbs can potentially be expressed as stored pockets of data that are released into the medium, either over time or by accident.

The Legacy of Tesla and the Pyramids

The connection between Tesla's work and the ancient pyramids is more than just a historical curiosity; it represents a continuity of knowledge and innovation that spans millennia. Both Tesla and the builders of many pyramids sought to harness the natural energy of the Earth to communicate with both the medium and the cosmos, using advanced materials and designs to create systems that could transmit energy and information across vast distances. As humanity continues to explore the potential of wireless energy transmission, the legacy of Tesla and the pyramids remains relevant. By combining ancient knowledge with modern technology, science may be able to achieve the dream of free, limitless energy for all, fulfilling the vision that both Tesla and the ancient Egyptians shared. However, for this to occur humanity needs to ensure that this technology does not fall into the hands of evil individuals. This aspect will always hinder the progression of the human species.

Notes

Chapter 5: You Can Hear Voices, But Don't Worry You Can Take Meditations For That

The medium is used by human beings and electronics in just about anything that emits energy, sound, or light. The human body, mind, and heart, when in unison, are all capable of utilizing the medium to communicate with others, manifest their own realities, and sometimes perform what is considered superhuman feats. Just as Wi-Fi uses the air to allow its waves to travel through it to transmit information invisibly, human beings do the same thing with each other—and in today's world when utilizing computation—in a similar way that light transmits information along its wavelength across billions of light-years.

During 2020, when the world was shut down due to COVID, many individuals who practiced meditation—whether visual or auditory—began to notice a significant shift in their experiences. Whether they were meditating in a field of grass, on an island with rushing waves, or on a peaceful, snow-covered mountain top, their meditative processes were interrupted and altered by external sources. It is these individuals who will understand this shift the most. However, four and a half years later, the majority of the population has come to recognize this phenomenon. As computers, AI, and other technological advancements provide more clarity of and through the medium, the reality remains that these technologies often fall into the hands of malicious individuals. These may include terrorist organizations, rogue government officials, and agencies that seek to exploit such tools to attack specific individuals—whether due to their drug use, anatomy, political and religious views, or even out of jealousy and broader/insidious goals.

When a person meditates, they place their brain in a specific hertz range. When in this range, the human mind can do things it normally cannot during everyday tasks. However, in the modern age—where humans constantly require the world around them to respond, react, and provide information at essentially light speed—meditation techniques have evolved. No longer is it necessary to sit in a lotus position, bring your hands or fingers together, slow down your breathing, and gradually place yourself within the medium. Now, people can achieve this state within fractions of a second, placing their mind in the medium and in that hertz range on command, and then utilizing it for their own benefit.

Malicious individuals may use this ability for their own gain, often at the expense of others' pain and displeasure. However, all it takes is a group of capable and kind individuals to counteract those using it for malicious purposes. The medium is something that people can actually perceive—not with their two eyes or by utilizing their occipital nerve, but in the same way one sees when they daydream. It's not just a trick of the mind; it can be manipulated by malicious individuals using anything from human talent to computers. But those who can place themselves in a meditative state within fractions of a second can see this field—they can see the medium and ride its waves.

They can combat those who use it against others; they cannot be beaten by typical computers, not even by supercomputers of today's world. However, technology will eventually advance to a point where all humans could fall victim to a computer or AI capable of achieving victory over the human mind. By then, the kind and positive individuals who have been fighting against this and refraining from using it against others' consent and will should have a contingency plan in place to combat the malicious mindsets, uses, and actions of those with an opposite charge. Eventually, even terrorist organizations will possess quantum supercomputers. Although the human mind is capable of defeating such technologies, malicious groups will still be able to disrupt the entire population at the blink of an eye. This means that all humans will need to be prepared to combat such groups in a timely, intellectual, and skillful manner. As a result, the human species as a whole needs to train themselves to migrate from the traditional lotus position to the light-speed reaction that some already utilize every day.

It is unfortunate that this knowledge was lost for decades, centuries, and millennia. However, the world is fortunate that it has technology it can rely on and utilize to expose this medium once more and use it for the benefit of the species as a whole. The key difference now is that every person must ensure that malicious individuals who use this medium to harass and harm others do not succeed, and that they become muted once more, as humans all once were.

The Medium is Also a Field of Communication

The medium, often referred to as the quantum field around us, is not just a vacuum where waves pass through; it is a complex and dynamic environment where various forms of energy, information, and communication occur. In this field light waves, sound waves, electromagnetic waves, and even quantum particles interact, creating a web of interconnected influences that shape reality as humanity at this moment perceives it. Human beings, whether they are aware of it or not, constantly interact with this medium. Our thoughts, emotions, and physical presence all contribute to and are influenced by the medium. In fact, the medium allows for instantaneous communication across vast distances, transcending the limitations of traditional communication methods. This communication isn't limited by the speed of light, as the medium itself provides a more profound and intrinsic connection between all matter and energy in the universe.

In scientific terms, this medium can be understood as the quantum field, a realm where particles are entangled, and information is exchanged at a microscopic level. The medium is not necessarily bound by the constraints of space and time, which makes it a unique and powerful tool for understanding the interconnectedness of all things. It is through this medium that light, energy, and information travel, impacts not just the physical world but the very fabric of reality itself.

Memory Lane has Oncoming Traffic

The medium can be manipulated by human consciousness and is, in fact, manipulated every day subconsciously. It holds all of the information around us, from memories of the past to precognition of future events. Even though the medium around us exists in our current time, it is not at all bound by time itself. The atoms and molecules that humans see and interact with daily are bound by the time produced by the gravity that warps space-time around us. However, the medium is not confined by that same timeframe or spatial scale. This is why certain individuals are capable of foresight. They can experience events that have not yet occurred, as well as recall events from the past without having lived through them.

As mentioned before, neurites are located in both the brain and the heart. There is a widely shared story from recent years about individuals who experienced a sudden attraction to foods, places, and mindsets after receiving a heart transplant. This phenomenon occurs because the heart contains over 40,000 neurites and neurons, much like the brain, which provides a connection to memories and experiences from the life of the individual who provided the heart. Science has shown that neurons and neurites do not hold data in the traditional sense. Instead, they act as antennas, communicating with other cells in the body as well as the medium around us. These neurons and neurites recollect past events, and can foresee the future when tuned into the medium in the correct manner. The well-known history of transplant recipients acquiring the donor's preferences, aversions, and even memories illustrates that these cells function as something other than just data storage. Neurons and neurites are, in essence, pulling information from the medium around us, including real-time events and future occurrences. This is not a novel concept—civilizations for centuries have made use of this ability, employing those who possess the attunement and awareness to access the medium's knowledge.

The human body is remarkably complex, and a variety of factors must align for someone to utilize such abilities effectively. It is relatively simple for the brain to recall past events or memories and convey them through speech, body language, or even art. However, when it comes to precognition and foresight, this skill requires everything to fall into place. The individual must have a full understanding of what they are doing, how to do it, when to do it, and what to seek out. Thousands of years ago, human civilizations would rely on individuals capable of traversing the medium and not only predicting future events but also reporting current ones from vast distances. These individuals were known as oracles or prophets. At times, these oracles and prophets would use their stature to manipulate and influence the civilization they served in devious ways. However, most often, they performed their duties as instructed, fulfilling their roles with fidelity. These prophetic individuals had the proper brain frequencies, and their cells were attuned to the medium around them, allowing them to traverse it outside the confines of time. For the most part, these prophets and oracles were drugged with narcotics, herbs, and other chemicals to help them access the medium. In today's world, however, humans are surrounded by so many frequencies and have evolved as a species to the point that it is much easier to traverse the medium without the use of any chemical compounds. Some of you reading this book may comprehend this easily, while others simply will not. To convince or teach an individual who cannot traverse the medium, and who has not attuned their body to it, is akin to telling a blind man there is a lake in front of him without allowing him to touch it or sense it in any way. Some blind men will believe you. Others simply will not, unless they touch the water for themselves. And even if the blind man touches the water, he may not truly understand what water is or what it looks like. Therefore, even if the blind man can feel the medium, he may not believe the person telling him that there is an entire lake in front of him. However, the blind man who believes in the lake without needing to touch it will be the one who can understand and reach the other side. He may walk around the lake, or sail across it with you. The medium is no different, and the blind man represents the vast majority of the population. This is why, when an individual foresees a future event, they will feel it, see it, hear it, and therefore know it. Those who cannot foresee or accept it fully will either remain on one side of the lake or join you on the other.

In ancient times, individuals who mastered these techniques were often revered as shamans, mystics, or prophets. Aside from the narcotic induced individuals, the truly foresighted ones had an innate ability to interact with the medium, receiving insights and information that were beyond the reach of ordinary people. These individuals were able to harness the power of the medium to heal, communicate with distant others, and even predict future events. As society evolved, the understanding and use of the medium became more structured and organized. Religious institutions, scientific communities, and even governments began to explore ways to control and manipulate the medium for various purposes. This led to the development of technologies that could artificially interact with the medium, such as radio waves, microwaves, and other forms of electromagnetic communication.

Today, the medium is not just a tool for spiritual or mystical practices; it is a fundamental aspect of modern technology. From the Wi-Fi signals that connect our devices to the quantum computers that promise to revolutionize computing, the medium is at the heart of our technological advancements. However, with this power comes responsibility. As humans continue to explore and exploit the medium, society must be mindful of the ethical implications and the potential for misuse.

Throughout history, humans have utilized various methods to tap into the medium, whether through meditation, prayer, or other spiritual practices. These methods were once considered esoteric and mystical, but modern science is beginning to uncover the mechanisms behind these practices. For instance, meditation has been shown to alter brainwave patterns, placing the mind in a state that is more receptive to the medium's influence.

The Happiest People in the World

Neuroscientist Richard Davidson at the University of Wisconsin-Madison conducted an extensive scientific study on a Tibetan Buddhist monk named Matthieu Ricard who was a close associate to the Dalai Lama. The research used EEG (electroencephalogram) technology to measure brain waves during meditation. Davidson sought to uncover the mysteries of meditation and its effects on the human mind. During the study, Davidson discovered that while Ricard was meditating, he produced an extremely high level of gamma waves—brain waves typically associated with intense focus, learning, memory, and feelings of happiness. The observed gamma oscillations were often above 100 Hz and were unusually strong and persistent compared to control subjects, which was a surprising result for Davidson and his research team. This suggested that Ricard's years of meditation practice had altered his brain's baseline activity, allowing him to experience heightened states of awareness and consciousness.

This scientific breakthrough highlights how meditation can enable the human brain to access frequencies and states of being previously thought unattainable without external stimulus. The human body and consciousness can manipulate the medium around it with proper training and techniques, and Davidson began to deduce this on a scientific level.

Davidson's study on Matthieu Ricard provided solid evidence that long-term meditation can induce neuroplastic changes in the brain. This means that, over time, meditation can actually alter the brain's structure and function. In Ricard's case, the regions of his brain tied to positive emotions were significantly more active than those in the control group. This suggests that meditation could enhance emotional regulation, increase overall mental well-being, and contribute to a lasting sense of happiness and empathy—something modern science is only starting to grasp.

Aside from the heightened gamma wave activity, Davidson's study also uncovered a notable increase in both alpha and theta waves in Ricard's brain. These brain waves are typically linked to deep relaxation, calmness, and intense focus. The high levels of alpha and theta waves further support the idea that meditation doesn't just relax the body—it brings about a profound state of inner calm and heightened awareness that defies the limitations of normal cognitive function.

Alpha waves are typically associated with calm wakefulness, the kind of state one enters when their mind is alert but relaxed. These waves surge when you're zoning out, drifting into thought while still awake. But what happens when you're actively training your mind to achieve that calm through meditation? What happens when you're not just zoning out, but harnessing that relaxation to deepen your connection with the medium around you? Ricard's brain, while meditating, produced heightened levels of alpha waves, which suggests that his mind was in a state of active, controlled relaxation. A state that most humans achieve only when their minds aimlessly drift. It's no wonder then that meditation can lead to heightened focus—Ricard's alpha waves were functioning at a level the vast majority of the population rarely taps into in daily life.

But then come the theta waves—typically linked with deep relaxation and light sleep. These waves are what you experience when you're teetering on the edge of unconsciousness. Theta waves allow for mental images, memories, and ideas to float into the conscious mind. The fact that Ricard produced elevated levels of these waves during meditation hints that meditation doesn't just relax the mind; it unlocks a door. It's the threshold between wakefulness and sleep where the mind is most pliable, most connected to the medium that holds memories of the past and visions of the future.

The surprising activity of both alpha and theta waves alongside the gamma waves—which are linked to heightened states of consciousness and happiness—revealed how deep meditation changes not just the state of mind but the fabric of brain function itself. The mind becomes a receiver, attuned to the waves around it, pulling from the medium in ways that are seemingly impossible for those not practiced in meditation.

This scientific breakthrough emphasizes the idea that meditation allows the brain to access states previously thought unreachable without external stimulus. It's not just about calming the mind; it's about tuning into something deeper, something innate within the human consciousness that connects us to the medium that holds all information—past, present, and future.

Davidson's study didn't just provide insight into the effects of meditation—it hinted at how the brain could be trained to access different frequencies, attune to the medium, and manipulate it. The increased gamma, alpha, and theta waves found in Ricard's brain serve as proof that the human brain, with the right training, is capable of traversing realms of consciousness and connecting to the medium in ways previously thought to belong only to myth and legend.

The Role of Non-Visible Light and Frequencies

While visible light is the most familiar form of electromagnetic radiation, the medium also encompasses a vast spectrum of non-visible light and frequencies, including ultraviolet light, infrared radiation, X-rays, and gamma rays. These forms of light carry information and energy that can have extreme effects on the medium and, by extension, on human beings and the world around us.

For example, gamma rays, which are a form of high-energy radiation, can penetrate deep into matter and alter its structure at a molecular level. This ability makes gamma rays both a powerful tool for medical treatments, such as cancer radiation therapy, and a potential hazard if not properly controlled. Similarly, infrared radiation is used in a variety of applications, from remote controls to thermal imaging, by exploiting its ability to transfer and observe heat energy through the medium.

In the scientific realm of quantum mechanics, non-visible light and frequencies play a critical role in certain phenomena such as quantum entanglement. Quantum entanglement occurs when particles become interconnected or "entangled" regardless of the distance between them. If one of the two particles is observed, the observed state of the first particle is expressed by the other particle. However, do not confuse this with a transfer of valuable data or information. It is essentially raw data describing the state or "spin" that the particle is in. It does occur at faster than light speed, and scientists are not entirely certain as to how or why this phenomena happens. This interconnectedness is facilitated by the medium, and expresses an instantaneous communication between entangled particles, a concept that challenges our traditional understanding of space and time. We will discuss quantum entanglement in a later chapter.

The medium also enables the transmission of information through these non-visible frequencies. In fact, many modern communication technologies rely on the ability to encode and transmit data using different frequencies of electromagnetic radiation. By manipulating the amplitude, frequency, or phase of these waves, vast amounts of information can be encoded and transmitted across great distances, effectively using the medium as a carrier for the digital world.

As science continues to explore the potential of non-visible light and frequencies, humanity will likely uncover new ways to harness the medium for both practical and theoretical purposes. However, with these discoveries comes the need for caution and responsibility, as the misuse of such powerful tools could have far-reaching consequences.

Voice to Skull (V2K) and the Modulation of the Medium

Voice to Skull (V2K) technology represents one of the more controversial and lesser-known uses of the medium. This technology enables the transmission of sound directly to a person's inner ear without the use of traditional auditory pathways. It does this by using modulated microwave frequencies to bypass the outer ear and eardrum, directly stimulating the auditory nerve and creating the perception of sound within the brain. Essentially, this means that a person can hear voices or sounds that seem to originate inside their own mind, despite the absence of any external source.

V2K technology was first developed in the 20th century and has been used in various capacities since then. During the Cold War, there were reports of the technology being used for psychological operations (psy-ops) and covert communication. Notably, during the Margaret Thatcher era in the United Kingdom, there were allegations that V2K was used to control riots and influence the behavior of certain individuals without their consent. This use of the technology sparked debates about the ethics and implications of such capabilities, as it raised concerns about privacy, mental autonomy, and the potential for misuse.

Today, V2K and similar technologies are employed in a range of applications, from military communication to law enforcement. In military contexts, V2K can be used for long-distance communication with soldiers behind enemy lines, especially in situations where traditional radio communication is not possible. This allows for the relay of critical information directly to the minds of soldiers without the need for physical devices that could be intercepted or compromised.

Law enforcement agencies have also explored the use of V2K for crowd control and riot management. By transmitting specific frequencies, they can potentially influence the behavior of large groups of people, calming tensions or dispersing crowds without the use of physical force. However, the use of V2K in this manner has been met with significant criticism due to the lack of transparency and the potential for abuse.

The technology's reach extends beyond just military and law enforcement; it has found its way into the hands of large corporations and individuals with enough resources to deploy it. These entities may use V2K to influence or manipulate specific targets, often without their knowledge or consent. The lack of regulation surrounding this technology only exacerbates the potential for misuse, raising questions about the ethical responsibilities of those who wield such power.

The effects of V2K on the medium and the broader electromagnetic environment are shown within all corners of society. By modulating microwave frequencies to transmit sound, V2K alters the natural state of the medium, introducing artificial signals that can interfere with other forms of communication and energy transfer. This has led to concerns about the long-term impact of such technologies on both human health and the environment.

Moreover, V2K technology is not just limited to auditory experiences; it can also be used to create visual hallucinations and other sensory experiences. By targeting different areas of the brain with modulated frequencies, it is possible to induce a range of perceptions, from seeing images that aren't there to feeling physical sensations that have no external cause. This capability has been explored for both therapeutic and manipulative purposes, further blurring the line between reality and artificially induced experiences.

The ethical implications of V2K technology are vast, particularly in a world where such capabilities are largely unregulated. While there are legitimate uses for V2K, especially in military and law enforcement contexts, the potential for abuse is significant. As scientific research continues to develop and deploy these technologies, it is crucial that there are established and clear guidelines or regulations to protect individuals from non-consensual and harmful uses.

Figure 1. The "Russian Woodpecker," a Cold War-era Soviet over-the-horizon radar system near Chernobyl, emitted a tapping sound that disrupted radio communications globally while serving as part of the USSR's missile defense early-warning system. It was dismantled after the Chernobyl disaster.

Notes

Chapter 6: You're Rather Bright, God Must Say

We are all beings of light. At our core, we come from light, we emit light, and we interact with light in ways that are fundamental to our existence. Every human being emits photons—tiny particles of light—along with energy and magnetism. This not only connects us to the cosmos but also to each other and the very fabric of reality. It is through light that we traverse the medium around us—the invisible fields, waves, and forces that permeate our world.

Navigating the Depths of Meditation: Reaching the Proper Humanistic and Spiritual State

Meditation is a journey inward, a deliberate effort to connect with the self, the universe, and the energies that bind us. Achieving the proper meditative state is not as simple as sitting down and closing your eyes; it requires discipline, knowledge, and an understanding of the body's energy centers, or chakras.

To begin, one must find a quiet space, free from distractions, where the mind can focus and the body can relax. The seven chakras play a crucial role in this process, serving as gateways to higher consciousness and deeper states of meditation. Each chakra, from the root to the crown, must be aligned and balanced to ensure that energy flows freely throughout the body.

The Seven Chakras and Their Role in Meditation

The seven chakras are the energy centers in our bodies where energy flows. Blocked energy in any of the chakras can lead to illness, so it's important to understand what each chakra represents and what we can do to keep this energy flowing freely.

Root Chakra (Muladhara):

Located at the base of the spine, it is associated with grounding and survival. It's the foundation of our physical and spiritual existence.

Sacral Chakra (Svadhisthana): Located just below the navel, it governs creativity, pleasure, and emotions.

Solar Plexus Chakra (Manipura): Located in the stomach area, it controls self-esteem, personal power, and willpower.

Heart Chakra (Anahata): Located in the very center of the chest, it is the source of love, compassion, and empathy. This chakra is particularly significant as it connects the lower chakras (associated with materialism) to the upper chakras (associated with spirituality).

Throat Chakra (Vishuddha): Located in the throat, it governs aspects of communication, self-expression, and truth.

Third Eye Chakra (Ajna): Located between the eyebrows, it is the center of intuition and foresight.

Crown Chakra (Sahasrara): Located at the top of the head, it is the gateway to higher consciousness and spiritual enlightenment.

When meditating, it's essential to focus on these chakras, visualizing them as spinning wheels of energy, each glowing with its respective color. The goal is to balance these chakras, allowing energy to flow freely from the base of the spine to the crown of the head.

Brainwaves and Meditation

Meditation alters the state of our brainwaves, transitioning them from the fast-paced beta waves, which are predominant when humans are awake and alert, to slower, more relaxed brain waves such as alpha, theta, and delta.

Beta Waves (13-30 Hz): These are associated with normal waking consciousness and a heightened state of alertness, logic, and critical reasoning. However, beta waves are also the waves of stress, anxiety, and restlessness.

Alpha Waves (8-12 Hz): These are present during deep relaxation with the eyes usually closed, when daydreaming, or during light meditation. Alpha waves bridge the conscious mind with the subconscious.

Theta Waves (4-8 Hz): Theta waves are the brainwaves of deep meditation and light sleep, including the all-important REM dream state. They are associated with creativity, intuition, and vivid imagery.

Delta Waves (0.5-4 Hz): These are the slowest brainwaves and are present in deep, dreamless sleep and in very deep meditation where awareness is fully detached. Delta is the realm of the unconscious mind.

In intense meditation, the brain predominantly produces theta waves, allowing for a deep state of meditation where deep-rooted inner work and healing can occur. Conversely, in light meditation, the brain shifts into alpha waves, a state of relaxed awareness that can be easily disrupted by external stimuli.

The Scientific Exploration of Meditation

Over the years, numerous scientific studies have explored the effects of meditation on the brain and body. Research involving experienced meditators, such as monks, has revealed significant differences in brain activity compared to non-meditators.

Functional MRI scans have shown that during deep meditation, the prefrontal cortex (associated with decision making and social behavior) is less active, while the posterior cingulate cortex (associated with self-reflection and mind-wandering) is more active. This shift in brain activity allows meditators to enter a state of heightened awareness and focus, often described as a state of flow.

Moreover, studies have shown that long-term meditation can lead to increased gray matter in the brain, particularly in areas associated with memory, empathy, and emotional regulation. These findings suggest that meditation not only changes the brain's activity but also its structure.

The Human Biofield and the Role of the Heart in Meditation

The human biofield, often referred to as the aura, is the electromagnetic field that surrounds and permeates the body. It is a reflection of our physical, emotional, and spiritual health. The heart, with its powerful electromagnetic field, plays a crucial role in maintaining the strength and coherence of the biofield.

The human heart generates the largest electromagnetic field in the body, which is about 60 times greater in amplitude than the brain's electrical activity. This field can be measured several feet away from the body and is influenced by our emotions. Positive emotions, such as love and compassion, produce a coherent heart rhythm, which enhances the biofield and strengthens our connection to the medium around us. Conversely, negative emotions, such as fear and anger, produce an incoherent heart rhythm, which weakens the biofield and makes us more susceptible to external influences.

This is expressed as a lack of brain-heart coherence. It is also dominated by the 40,000 neurons and neuritis in the heart, which channel memories and can trigger strong emotions. They are the antennae which channel the medium around us, and can produce visual and auditory memories due to trauma or incredible experiences that both the brain and heart experience. Heart transplant patients will express this to you with their sudden desire for things, and memories of, experiences they have never had. This is well documented and acknowledged by both scientists, doctors, and spiritual teachers alike. Heart transplant patients sometimes have cravings for foods, activities, environments, and vices that they were never exposed to prior to their surgeries. Some even behind memories, that provide visual and auditory experiences of the memories, that they never experienced or possessed prior to the surgery. This is why traversing the medium with strength and clarity can only be achieved by those that abide by the coherence of the brain and heart—which is dictated by both the truths and positivity of the medium and their inner self.

In deep meditation, the heart's coherent rhythm enhances the biofield, creating a state of harmony between the mind, body, and spirit. This state allows for a more profound meditative experience, where the individual can traverse the medium, access higher states of consciousness, and connect with the universal energies.

The Encoding of Bacteria that Transcended Deca-Generationally

Recently, scientists conducted research and experiments focused on encoding the DNA of bacteria with specific data and information. This encoded data successfully transcended over twenty generations of the bacteria without any decoherence of information. Researchers at Harvard University successfully encoded data into the DNA of the bacterium Escherichia coli (E. coli). The team used CRISPR-Cas9, a precise gene-editing tool, to insert digital information—such as text, images, and even short videos—into the bacterial genome. This data was stored within the DNA sequences and maintained its integrity as the bacteria replicated over multiple generations. Remarkably, the encoded information was preserved with high fidelity, demonstrating the potential for DNA as a long-term data storage medium. The experiment began by converting binary data into nucleotide sequences, which were then integrated into the bacterial DNA at specific loci. Each bacterial cell effectively acted as a "biological hard drive," capable of passing on the encoded data through the natural process of replication. The researchers used a process called "recombinase-mediated cassette exchange" to insert the data segments into the bacterial genome, ensuring that the encoded information would not disrupt essential genes or cellular functions. To retrieve the stored information, the researchers extracted the bacterial DNA and sequenced it. They then decoded the nucleotide sequences back into binary data, recovering the original information with over 90% accuracy. This experiment not only demonstrated the feasibility of using bacterial DNA for data storage but also opened new avenues for bioinformatics and the intersection of biology and information technology.

This is one example of scientifically proven methods in which DNA and genetics can be encoded with information and data that maintain their integrity across multiple generations. This information extends beyond simple physical attributes, chemical compositions, and biological functions. A notable case by Dr. Paul illustrating the profound capacity of biological encoding is that of a young girl who, after receiving a heart transplant from another girl who was murdered, began experiencing vivid nightmares and visions of the donor's murder. These experiences were so detailed and consistent that law enforcement was able to use them to identify and convict the murderer, based on the information provided by the recipient's memories. It is now understood that neurons and neurites do not store memories themselves but act as antennas to access memories, form new ones, and interact with the medium in which information and memories are stored. This antenna-like function also communicates with DNA and genetic material encoded within the cells and throughout the biological form it inhabits. Given that information within DNA can be transmitted across generations, it is conceivable that memories and specific moments in time, embedded within a lineage's genetic code, can be accessed. This can be achieved through a mediator, such as a living descendant, or even through individuals who receive DNA encoded with the information needed to recall the memories of a person they wish to examine, like organ transplant recipients.

The Role of Emotions in Meditation and the Power of the Heart

While the mind plays a significant role in meditation, it is ultimately the heart that dictates the strength and depth of the meditative experience. The heart is the seat of emotions, and it is through the emotions that we connect with the universal energies.

Emotional intelligence, the ability to recognize, understand, and manage our emotions, is crucial in meditation. A person with high emotional intelligence can enter a meditative state more easily, maintain focus, and navigate the challenges that arise during meditation. However, it is not enough to be emotionally intelligent; one must also be emotionally mature, capable of processing and releasing negative emotions that can disrupt the meditative state.

The heart's electromagnetic field is directly influenced by our emotions. Positive emotions, such as love, gratitude, and compassion, strengthen the field, while negative emotions weaken it. Therefore, the strength of your meditation is not just determined by your ability to focus or quiet the mind, but by the emotions you bring into the meditative state. Coherence and truthfulness regardless of intention and overcoming personal deceit are required to reach true brain-heart coherence.

I will not explain how to cohere your brain and heart—it is a deeply personal aspect of being human. However, it is essential to reach a heightened level of meditation, strength, and the ability to traverse the medium effectively and for great lengths. Those who understand brain and heart coherence profoundly will grasp what they have just read. For those who do not, the journey begins by asking themselves a simple question—not aloud, but internally. Channel your heart and your mind while asking the question, and don't defend against emotions from either side. Injustice will be felt on both ends of the spectrum. When injustice is felt against your own beliefs and mindsets, it serves as a signal. Regardless of how evil or malicious you may be in the situation, your heart will stop you. Conversely, no matter how unbalanced the situation may be against your favor due to injustice, your heart will never let you lose. When you reach a complete understanding of yourself—your emotions, maturity, and intellect—you will be gifted with everything the medium can provide to aid you in your pursuit of your desires. This is why not everyone is seen walking around, instantly flashing into a meditative state, accessing the medium, and acting at will. Only those who are worthy can achieve this.

Otherwise, you will need to take the time to lower your heart rate, slow your breathing, produce melatonin, reach the proper brainwaves, and then elevate your brainwaves to access the medium. In the time it took to read that sentence, those who are truly capable have already been traversing the medium by the third word. Until then, know yourself, know your heart, know what is right, and know what is wrong. Balance the situation based on the injustices that humans all innately comprehend at a spiritual level. Do not confuse this with visual and auditory experiences shared like a group daydream or meditation. This is something much more powerful and profound, and is the baseline for those that break through the glass ceiling of being human. Remember, external influence from malicious others doing the same as you may lead you astray if they are under the influence and can comprehend this aspect of meditation and medium-traversing. Most drugs can induce higher states of consciousness through higher frequency brainwaves, creating individuals that lack coherence but reside within the medium. However, these individuals are easy to defeat by acknowledging this sole fact and causing their hearts and emotions to disrupt their brain wavefunctions.

The Hiccups Heard Around the Psychological World: The Role of Hiccuping in Detecting External Influences

Pulse modulated microwaves have been studied extensively for their potential to alter brain functions, expressions, and states. This technology manipulates brain wave frequencies, creating a "back and forth" communication within the brain that can be described similarly to binaural beats. By observing the resulting frequency, derived from the median integer of an initial frequency that resides in the 100 MHz - 210 MHz range, and the brain's natural brain frequency. Researchers gain valuable insight into the internal functions and thought processes of the targeted individual. Within this context, hiccuping stands out as an innate and automatic bodily reaction, providing a unique form of data to distinguish between external inputs and genuine bodily signals. This involuntary action, which is nearly impossible to predict or control on command, serves as a biological flag that highlights any deviations caused by external influences.

When the brain is monitored during a hiccuping phase, the data collected offers a clear distinction between external signals and those produced organically by the body. The involuntary nature of hiccuping makes it a reliable sentinel, signaling discrepancies that might be masked during other physiological activities. This becomes especially vital in scenarios of electronic harassment or targeting, where the aim is to manipulate or disrupt an individual's mental state. The hiccup acts as a natural and unbreakable defense mechanism, revealing inconsistencies in brain wave patterns that could indicate external tampering. While it may seem like an inconsequential reaction, the hiccup provides a robust point of reference for decoding external interference and determining the authenticity of neurological data. In the larger scheme of counteracting electronic harassment, understanding and utilizing these innate biological responses could prove essential in identifying and mitigating the negative effects imposed on targeted individuals. Though seemingly trivial, hiccuping may well be a key player in the endgame against those who seek to manipulate and harm through technological means.

This automatic and uncontrollable action by the body provides individuals reading brain waves and functions a key bookmark for what the actual body intends to perform and think, compared to the external input applied to the individual. The external input can at times be seamless and virtually undetectable unless an inadvertent and uncontrollable action, like a hiccup, occurs. I experienced this, of course, inadvertently one night while writing this book. While being monitored under security clearance both remotely and closely, I began to hiccup. This triggered a response from those that were influencing and monitoring my brain waves and functions. This allowed the individuals that were monitoring me for external input to differentiate between my own thoughts or actions and those implementing thoughts and actions. Once this occurred, a flurry of reactions from those pretending to be God, myself, and various other people began to unfold. These reactions were reflected by my body in numerous manners, however, the importance of this occurrence superseded their resulting reactions. Eventually, all external inputs were flagged and the study being performed on me was easily able to predict and depict the subtle differences between my free will and those imposing even slight movements upon me. I wish I could say there is an easy way to repeat this type of check and balance, but it seems as though a bodily function and inadvertent actions like the hiccup is crucial when dissecting the differences between external forces being applied and true internal actions.

Mother's Know Best, But They Can Thank Their Brother

In the modern understanding of genetics it is commonly understood that only 2% of the human genome has been fully decoded, leaving 98% in a realm of ambiguity. This "junk DNA" is anything but useless; it's a vast, unexplored territory that may hold the key to understanding complex phenomena like memory transfer through organ transplants, reaction to specific stimuli, memories of predecessors and living relatives encoded within the medium, and microchimerism between siblings. The stories of transplant recipients experiencing the donor's emotions, cravings, and even memories provide a unique perspective into a theory that, while not yet scientifically validated, cannot be dismissed.

Let's consider the case of organ transplants. There are numerous accounts where recipients adopt preferences or traits of their donors post-surgery. These experiences suggest that there is more to the cells and DNA in organs than just their biological function. While mainstream science has yet to pinpoint a molecular mechanism to explain how memories or experiences could be transferred, the persistence of these stories indicates that something is happening at the cellular level that we don't yet understand.

This phenomenon doesn't stop with organ transplants. Enter microchimerism—a process where cells from one individual remain in another's body long after pregnancy. In cases where a mother has multiple children, cells from the first child can be present in her body and then be transferred to subsequent children. For example, a first-born daughter could leave a cellular imprint that is later shared with her younger brothers. This cellular presence is not about direct genetic inheritance but rather about shared biological history and potential memory imprints.

Now, let's take this a step further. If these shared cells contain more than just genetic information—if they somehow carry a trace of experience or emotional resonance—then siblings could, in a way, be connected on a level deeper than previously thought. Imagine the implications: two younger brothers, carrying cells from their older sister, might have a subtle, imperceptible link to her life experiences. This link could, theoretically, create a form of emotional or intuitive awareness between them, extending even to the older sister's children.

This dynamic becomes even more intriguing when considering the order of birth and the natural connections that may arise between siblings. The older sister, having shared her cellular makeup with her younger brothers, establishes a unique bond that could act as a conduit for emotional and intuitive signals. Just as a mother can intuitively sense when her child is in danger, this cellular bond might extend the sister's maternal intuition to her brothers. If the older sister experiences a strong feeling about her own children, her younger brothers, connected through shared cellular memory, could potentially sense the dangers or joys that her offspring—and she herself—experience on a regular basis. This suggests that the bond of microchimerism and familial connection might serve as an invisible thread, linking siblings and allowing them to resonate with each other's emotions and experiences in ways we are only beginning to understand. While not a direct genetic inheritance, this could be a form of cellular empathy, an echo of shared biological threads that link them to their niece or nephew in unexpected ways.

The idea that a mother can instinctively sense when her child is in danger is well-documented. There is no doubt that this intuition extends to siblings. Because if a mother's cells are shared among her children, and these cells carry information beyond mere biology, then her sense of connectedness could ripple through her offspring. When a mother feels her soldier son's distress before his brigade knows he's in danger, it may not be just maternal instinct. It could be that the cells she shares with him—and possibly with her other children—are responding in ways we don't yet comprehend.

This isn't about asserting that memories and experiences are transferred in a literal sense through DNA or that cells function like miniature hard drives. It's about recognizing that the human body—and by extension, the human experience in the medium around us—is far more interconnected than our current scientific models can fully explain. The persistence of these phenomena, whether through organ transplants or microchimerism, suggests that our understanding of consciousness, memory, and familial bonds is still in its infancy.

We can't dismiss these stories just because they don't fit neatly into established frameworks. Remember, we understand 2% of our own DNA. Instead, we should view them as clues pointing toward a more complex reality. As science advances, it may reveal that the so-called "junk DNA" is not junk at all but a repository of information that plays a crucial role in shaping our identities, our connections to each other, and the world around us.

While we lack concrete experimental evidence to prove these theories, common sense and logical analysis suggest that there is more at play here than we currently comprehend. I personally have experienced similar instances, therefore, am one of many that prove this aspect of my theory to be true. The stories of transplant recipients and the concept of microchimerism offer a glimpse into a medium where our bodies are not isolated entities but parts of a vast network of biological and possibly even experiential connections. This perspective demands that we remain open to possibilities and continue to explore the mysteries of the human genome and beyond.

The connection between neurons and neurites with the surrounding medium offers a profound insight into how real-time experiences, particularly trauma—whether physical, emotional, spiritual, or mental—might be shared among family members. Neurons and neurites function not as storage units for memories, but as transmitters and receivers that interact dynamically with the medium that encompasses all things. Acting like antennae, they can tune into the energetic and informational imprints embedded in this medium, which includes the real-time occurrences of trauma and intense emotional experiences.

When a family is biologically connected through shared DNA and cellular components, this bond might facilitate a heightened sensitivity to the traumas experienced by one another. If a family member undergoes significant physical pain or emotional distress, the shared biological resonance could allow neurons and neurites in other family members to pick up on these disturbances through the medium. This means that, beyond mere empathy, there could be an actual transference or resonance of the trauma itself—whether it's a physical sensation, a wave of emotional anguish, or a spiritual burden. The family's collective connection to the medium could enable a profound, real-time exchange of these intense experiences, making them more than just bystanders to each other's suffering, but active participants in a shared biological and energetic response.

This profound connection between family members is not just an accidental phenomenon but an innate quality of the evolutionary development of DNA, or by the design of God, within each family's genetics. The ability to sense and respond to the trauma of one another in real-time—whether physical pain, emotional distress, or spiritual turmoil—serves a fundamental purpose: the survival and success of the family lineage. By being acutely attuned to the experiences of one another, family members can react swiftly to threats, provide support, and adapt to challenges collectively, ensuring the well-being and resilience of the entire family. It also allows future generations to receive a modified version of the DNA, which provides an innate response to certain complex circumstances that has plagued or damaged previous generations within their family's lineage. This would transpire through genetic alterations between relatives that are closely related, or share a passed down microchimerism cell or DNA strand.

This heightened sensitivity to each other's states can be seen as an evolutionary mechanism embedded within our DNA, crafted over countless generations. It is a sophisticated system of shared awareness that equips family members, particularly offspring, with the necessary traits, knowledge, and instincts for survival and reproduction. The transmission of not only physical responses but also emotional and mental cues helps prepare the younger generation for the complex realities they will face, enhancing their ability to thrive and propagate the lineage. For example: if the oldest sibling has two offspring, and the oldest sibling's younger brothers experience specific traumatic experiences that result in near death experiences or death itself due to genetically specific horrors such as genocide, then it would be not only imperative but innately essential for the successors of the family, regardless of parent, to pass this information along and modify the successors DNA and how they respond and acknowledge to said dangers. To think that the human genome isn't complex enough to allow this to occur, and that the medium doesn't play a role in mediating this, is illogical. This interwoven network of biological and energetic connections acts as a living legacy, passing down not just genetic material but a deeply rooted, dynamic support system that ensures the continuity and success of future generations.

The Influence of Malicious Input on Meditation and the Medium

In the modern world, the world is constantly bombarded by external stimuli, some of which are deliberately designed to disrupt our mental and emotional balance. During meditation, these external influences can manifest as intrusive thoughts, uncomfortable sensations, or even false perceptions. This is a saddening section of the book, but must be spoken about if you are to comprehend the medium and all that reside within it.

Malicious individuals or entities can exploit this vulnerability, using technology or psychological manipulation to interfere with your meditation. This interference can take many forms, from subtle distractions that break your focus to more invasive techniques that distort your sensory experiences. These external stimuli can be anything from V2K technologies, to simple sound frequencies in a specific hertz range, to another individual under the influence of narcotics or medication that uses what the chemical compound does to their brainwave frequencies to their advantage and your disadvantage. Both computation and drug influence can place someone in the medium, with which they can harass, disrupt, or torment the recipient. This is nothing new to meditators that are well versed in the field of the medium. However, all are easily defeated.

To do so, it's essential to develop a strong awareness of your own mind and emotions, so you can recognize when an external influence is attempting to disrupt your meditation. By maintaining a coherent heart rhythm and a strong biofield, you can protect yourself from these intrusions and maintain the integrity of your meditative experience. Once you understand yourself, tune yourself into what the malicious individual is attempting to either project or instill within you. This will provide insight as to why they are there harassing, tormenting, or disrupting your meditative process of medium walking. Once you acknowledge their desires, you can combat them with the ethical and moral dilemmas they are encountering by doing such harmful acts. Over time, this will damage them, regardless of if they continue to pursue their disruptive behaviors. Eventually, if an individual spends every day disrupting your meditation, or disrupting your life through meditation and the medium, and you are consistent with cohering their hearts with and brain with inequalities and injustices, they will have a hard time advancing any further in the realm of the medium and meditation. You can actually train the harasser's heart and brain to permanently lack coherence when doing so, causing trauma to the attacker. So that every time they try to advance into the medium they experience a traumatic experience and eventually lack enjoyment or fulfillment from it. That permanent effect will stay with them forever, hindering them from the advancements that both the scientific field and spiritual world will and have uncovered. I will not teach such things, but those that have been harassed, abused, raped, disrupted, or harmed through the medium know what I am speaking about. Contact me, A.J., if contacting the authorities does not suffice. I am here for you. Those that experience disruptions, abuse, or harm during sleep are no different than those ill equipped from such malicious acts when they are awake. I have been a victim of horrific acts both asleep and awake. Eventually, I was able to overcome all obstacles when awake and most obstacles when asleep. Lack of REM can help break their cycle. But dream walking is something that can be difficult to overcome without proper lawful action. When you are asleep your brain will eventually migrate into a specific frequency range that exposes you to the malicious individuals you encounter when you are awake. It is important to understand that at times there is nothing you can do while asleep. When you are asleep, and an individual chooses to use V2K or pulse modulated microwaves to control or influence your dream state, it is difficult to combat these malicious individuals when unconscious. This is not a lack of strength, ability, or anything of the sort. These are normally weak minded, feeble individuals that are most likely already broken by you or someone else. Anyone who is molested while asleep isn't the loser, it is the loser that molests someone during their sleep because they need the advantage to do so. They need to feel as though they have power over someone. No, they are not the devil or demons. I promise you.

One method I have used while asleep to discover that certain things were occuring, and to confirm that external input was being applied to me while I was asleep, was to simply-as odd as it may seem-zip tie my hands and wrists to my bed frame. When I started to do that, I initially woke up with them still on, untouched. Eventually after a couple of days, I began to wake up with the zip ties undone, loosened, removed, or cut. At one point, I would zip tie my wrists so tight that it was uncomfortable and would be close to cutting off circulation, just to ensure that they couldn't be removed so easily. But eventually, as I continued to do that, I would wake up with either cuts on my hands, both zip ties cut off, or one zip tie undone while the other was still attached. This proved to me and countless others watching over me that something, whether it be choreographed by computation or driven by other human beings, was performing disgusting acts to me and others, as well as my nieces, while we are asleep. As I said earlier, my nieces and I have experienced similar things, both awake and asleep. I have stayed over their house and awoken to horrors that were applied to both me and them. However, I woke up to them screaming and crying, with my sister not knowing what to do. My nieces, who were 4 and 6 years old at the time, have even drawn some of the things they and I have experienced and seen while awake and asleep. At one point in my life I stayed awake for days in a row, waiting for the disgusting input to arrive so that I can defend both them and I. I am someone who understands them, and although many readers may not, there are countless others that will read this book and understand what I am speaking about. There are individuals that traverse and utilize the medium, not in godly or divine ways, but in evil and malicious ways. They are broken souls, stuck in an area of the medium with no chance of escaping.

So my suggestion to those that experience or think they may experience something similar is: to utilize what democracy has utilized for hundreds of years. Checks and balances. If you feel as though you are experiencing inadvertent movements, behaviors, or other various inexplicable horrors while asleep or awake, simply zip tie your wrists (tightly but not too tight) right before you fall asleep. When you wake up, see if they are still attached. Do this for a few weeks, you won't mind because YOU are asleep. When you wake up in the morning you can undo your zip ties and go on about your day. If you wake up in the morning and the zip ties aren't attached to your wrists, you know for a fact that there are individuals that are harassing you in your sleep, utilizing your body in ways that I don't want to describe. It is one of the easiest checks and balances to perform, since you are not awake for the majority of the experiment. It costs the least amount of money, and it will motivate the individual to speak to authorities, parents, and guardians with substantial evidence. I personally have woken up in the morning with cuts from my scissors all over my hands, with the zip tie either removed or not, clearly showing attempts to cut off the zip ties while asleep. This means the individuals that are harassing me, and others, aren't accurate enough to have complete and clerical control over the human body. This is not Godly, nor divine in any manner. They cannot perform these tasks in ways that a surgeon can simply perform their duties with a scalpel, a mechanic with a screwdriver unscrewing a screw, or an artist drawing a work of art. Therefore, they are untalented and are in the bottom end of their class (what a terrible school). This means they can easily be defeated, especially by those that are professionals in the field.

This is no different than those that fornicate with the deceased—take advantage of another when unconscious to achieve the feeling of empowerment. I have experienced, and my very young nieces have experienced—and unfortunately they still experience—such horrors while at rest. It is both from computation and disgusting individuals. While it is important to contact the authorities, at times they are ill equipped or lack the will to do something about this. Having someone alternate taking night shifts, keeping watch over a period of time can hinder, eventually stop, and eventually catch those individuals. Harassers, molesters, and abusers know their days are numbered, and typically do not quit until they are caught. Stay diligent if law enforcement cannot resolve the issue for you swiftly, but do know there are departments in law enforcement that handle these types of scenarios. I know individuals that have created groups with members that live all around the world to combat such disgusting acts, and it works if there is at least one person up at all times. The ability to remote view, or flag the abuser through technological means or human ability can speed up this process. Reasoning with such individuals is futile, just like reasoning with someone who abused or rapes an unconscious person, or a cadaver, can't be reasoned with. They must be found, accounted for, and processed (or reprocessed) through the judicial system accordingly.

Navigating the Medium with Awareness and Protection

Traversing the medium during meditation requires not only focus and emotional strength but also vigilance. The medium is a vast, interconnected field of energy, and while it holds the potential for necessary spiritual experiences, it also contains dangers.

To protect yourself while meditating, it's important to establish a clear intention and create a protective field around yourself. This can be done by visualizing a shield of light surrounding your body, or by focusing on the heart chakra and generating a strong, coherent rhythm that extends into your biofield.

As you navigate the medium, be mindful of any changes in your thoughts, emotions, or physical sensations. If you notice anything unusual or uncomfortable, take a moment to ground yourself, focus on your breath, and reinforce your protective field. Remember, you have the power to control your meditative experience and to reject any external influences that do not align with your intentions.

By understanding the interplay between the chakras, brainwaves, emotions, and the biofield, you can deepen your meditation practice and harness the full potential of the medium. As you continue on your journey, remember that meditation is not just a tool for relaxation, but a powerful means of connecting with the universal energies that shape our existence.

The Armor of Mjolnir, the Power of a Lion, the Heart of Ophiuchus, the Conceptual Justice of a Libra, and the Awakening of a Cancri

The ability to place yourself, your brainwaves, and your chemical makeup in a meditative state far quicker than it takes to cross your legs requires a profound understanding of the medium around you, a mature understanding of your heart's intelligence, your brain and heart coherence, and the sheer determination in the situation you're in to achieve this state of mind. For example, everybody has heard the story of the mother whose baby is trapped underneath a car. In that moment, the mother suddenly, within seconds, can lift a quarter to a half of a car's weight with just her strength for a few seconds to pull the baby out from underneath the vehicle and save its life. Some say it's a primal instinct, an instinctual reaction, that all humans can innately produce this determined superhuman feat when the time comes. It is those who tap into that aspect of being human that can place themselves in a strong, powerful, and effective meditative state at the blink of an eye.

It is correct in saying that all humans can innately perform this task when in the proper situation of distress. However, you do not need to be in distress to do this. You need to train yourself mentally, emotionally, and physically to tap into how your body performs this superhuman feat when in distress, so that you can use it to your advantage at any time. This requires the disciplined development of emotional intelligence, where your heart's energy aligns with your mental focus, creating a unified force capable of navigating the medium with precision and strength.

Through consistent practice and the conscious synchronization of your heart and mind, you can command your internal energies to react instantaneously, just as a mother's instincts kick in to save her child. By cultivating this skill, you gain the ability to enter a meditative state rapidly, allowing you to traverse the medium and harness the energy around you to enhance your experiences, protect yourself from malicious influences, and achieve a heightened state of awareness. This state of awareness is not just about inner peace, but about empowering yourself to influence the medium, react to external forces with resilience, and maintain control over your environment and well-being. Being able to achieve this state of mind quickly and effectively, and sustaining it for longer than a few moments, requires your brain and heart to be in full coherence with the reason you want to reach this meditative state so rapidly. You can fool the brain very, very easily with visual and auditory experiences, as well as trickery with reasoning and deceit. However, you cannot fool the heart, which is key in an advanced meditative state. When the reason you want to reach a meditative state quickly and abruptly is approved by your mind, body, and therefore your soul, you will achieve it. Just like the mother picking up the car, nothing can stop her because the reason she needed that strength affected no one negatively or adversely at all—it could only save a life.

Therefore, if one does not practice emotional maturity and heart-brain coherence, one will not achieve the most powerful meditative state possible. However, if you are fair to yourself and others, and deep down, no matter how devious you are, you know the reason you need to enter a meditative state so abruptly and so quickly is advantageous not just for yourself but for everyone around you in terms of protection, progression, and justice, then it is easily achieved. Do not confuse this with rage. Do not confuse this with anger. Do not confuse this with brute force. This is taking your brainwaves from a resting rate of anywhere between 8 hertz and 30 hertz, dropping it down, and then raising it to 30-100 hertz within a second or two. This requires your body to communicate with all the aspects needed to reach that state of mind quickly and to pass all the checks and balances it asks, which even the most devious and evil individual will find hard to do because somewhere out of the 40,000 neurites in their heart, something will stop it. Those 40,000 neurites are not just developed by the human brain's opinion and the human heart's agony; they are attuned to the world around you and therefore will stop you if it is not deemed correct, balanced, or safe.

This state of mind coincides with the fight-or-flight mentality, an instinctual protective mechanism for human beings. It requires both the brain and the heart, leading to a rise in both brain waves and energy. This is why those who experience fight responses, even when they should not, enter this heightened state of mind and physiological state for a few moments. When everything is in balance, this state can be sustained, but it is the latter—the extended and balanced state—that enables superhuman feats, while the former only provides a brief taste of it. It is why Jesus Christ was able to endure the cross for so long, and why legends like King Leonidas were able to fight off thousands with only hundreds. Individuals who face extreme injustice often overcome all obstacles not just because of sheer human will, desire, and tenacity—though these play a part—but because of their connection to the medium, communicated through their energy and expressed through their minds, hearts, and physiological being into the world around them.

The Science of Light and Its Connection to Life

Light is more than just a wave or a particle—it is a fundamental force that shapes the universe and everything within it. The speed of light, at approximately 299,792,458 meters per second, is the ultimate speed limit of the cosmos, governing the laws of physics and the flow of time. But light is also a carrier of information, encoded with the history of the stars, the composition of matter, and the interactions of energy and force.

In the human body, light plays a critical role in regulating biological processes. Photons interact with our cells, influencing everything from our circadian rhythms to our mental states. The pineal gland, for example, responds to light by producing melatonin, a hormone that regulates sleep. The production of melatonin aids in the ability to meditate more thoroughly and easily. Melatonin brings the brain waves to a specific hertz, which aids the person in entering the medium with which it can traverse. Yes, it can make you fall asleep, but this specific brain wavelength is crucial for the beginning of the meditative state. From there, the mind, body, and soul, can work in unison to achieve the goals with which those that meditate the strongest and most profoundly reside. It is the spark, or initial strike the brain needs to quickly and efficiently enter a different state of mind. But beyond its biological functions, light also has a more mystical aspect—it is the medium through which we connect with higher realms of consciousness.

Photons are not just particles of energy; they are also messengers, carrying information across vast distances. When light from a distant star reaches Earth, it brings with it a record of the star's history, encoded in its wavelength and frequency. Similarly, the light emitted by our bodies carries information about our physical and emotional states, creating a field of energy that interacts with the world around us. This is why external input can sometimes positively affect a person's mental state, emotional state, and physiological being. Many artists, actors, directors, scientists, and influential figures throughout history have stated that their works of art, research, and knowledge do not always come from them but through them. But one must ask where that information originates to flow through them. This brings up the point of the divinity of light, frequencies, and the medium that surrounds us. Light carries information, but it is not just data about how hot a star is, how far away it is, or what it encountered on its way to us. Just as today's technology encodes light with information to communicate quickly and easily, the frequencies and light around us are encoded in ways that enable them to speak through us.

This encoding is a necessary part of the checks and balances that your mind, heart, body, and soul must undergo for true brain and heart coherence, allowing you to traverse the medium in ways that may seem like science fiction or superhuman feats. Just as select artists and influential figures have had their information, which has influenced humanity, speak through them, you need to be—or become—an individual with comprehension of the medium around you, justice toward others, coherence between your brain and heart, and a genuinely kind nature to allow the divinity of light and frequencies to speak through you. This requires truthfulness to oneself and others, and the ability to observe the positive progression of the world around you as a result of the influence you provide to the medium.

To Be Immaculate Is Not a Requirement for Divine Enlightenment, But the Teachings of it Certainly Are

Many influential figures throughout history have spoken about ideas, creativity, and inspiration flowing through them. Some wonder where this inspiration comes from, and it cannot be coincidence:

"I don't believe I have great musical talent. I just had a lot of inspiration and learned to be able to let that inspiration flow through me" & "I'm just a channel. Everything comes to me. Everything flows through me." — John Lennon.

"Creativity requires faith. Faith requires that we relinquish control. This is frightening, and we resist. The truth is that it is far easier to get ourselves out of the way in order to let God work through us." — Julia Cameron.

"When the creative muse comes, you have to be open to it. My job is to stay out of its way and just let it flow." — Stevie Nicks

"I saw the angel in the marble and carved until I set him free." — Michelangelo

All the above quotes require an individual to strip oneself of ego, have faith in the process, seek positive influence with their work, and channel the medium around them to receive this information. Divinity of light and frequencies does not require one to have immaculate conceptions, but the mindsets of Jesus Christ and figures alike seems almost like a requirement for the attunement of this type of true enlightenment and influence. For God and the medium around us knows all, and will react to you accordingly.

Hollywood has depicted our contact with other intelligent species as a scenario driven by violence, power, destruction, and either extermination or colonization. However, I do not think this is the case. I believe truly intelligent species and beings—whether they be angelic, whether they be God himself, or just another species from another galaxy—must inherently be nonviolent, empathetic by nature, and cooperative. They would understand that life itself is rare and that working as a team generally leads to success for everyone involved.

These are the teachings of many religious, historical, and spiritual figures, as well as mantras that countless people still live by today. Individuals who embody these principles are always remembered, revered, and immortalized through stories, literature, and memory throughout the history of humanity.

The Dual Correlation of Light, the Æon of Pisces, and How to Form Slits in the Great Filter'sWall

There is undoubtedly a correlation between the duality of the aeon of Pisces that we currently exist in here on earth, and the duality of all light—the shining light in which we worship, share, and use to transmit information. It is a hint at the type of reality we are currently bound to biologically on this planet. It is when you take the duality of Pisces and push it through many slits you see the differences between the obvious and tangible aspects of Pisces and the broader, limitless spectrum of what it has to provide. Only then, will an individual fully comprehend that statement.

Let's discuss how one would create a double-slit experiment for the duality of Pisces and it's æon's spiritual teacher. The "particle" for Pisces—not referring to the light emitted from the constellation—is indefinitely encoded with information and data that won't be fully comprehended without the ability to dissect the information and language encoded within its entirety. The "particle" of Pisces would essentially be everything that we can easily see. Pisces is associated with themes of spirituality, belief systems, and dualism. But Pisces also correlates with the æon that we currently exist in and will continue to exist in for hundreds more years. It is inextricably tied to Jesus Christ, considering his dawn of existence at the beginning of this age, his double edged sword that appears in his mouth, his control over death and Hades, and his Father who resides and controls the heavens. The particle can be viewed as the direct story and images of him—even though photography did not exist back then, and depictions of Jesus in paintings and scriptures didn't appear until hundreds of years after his death. These literal, tangible objects would be considered the particle of Pisces, both the æon and bodily form...if there is even a difference.

The waveform of Pisces is what you need to filter and bring to light in order to comprehend its duality. This means you need to filter and dissect the light. Without doing so, one can never truly comprehend and understand what the æon has to offer—in a physical and literal sense, and in a spiritual and ethereal sense. But what do we use to create slits in the barrier that provides the interference pattern which is between the particle and the wave, allowing us to comprehend what is being communicated by the divine nature of the æon and the cosmos that surrounds us?

Let's define the barrier that needs the slits for what it truly is: a barrier that blocks the entirety of the light. If it has no slits to cause an interference pattern, it's just another wall that blocks the light from reaching its destination. Of course, I'm not speaking about literal light from Pisces or the stars that reside within it—though if we truly comprehended the data encoded within them, we would find important and quicker answers. But until we learn to speak the proper language, this is more of a metaphorical and spiritual sense. Therefore, the wall that blocks the light and the truth of the æon we live in must be something both internal and external. Something strong enough to prevent us from seeing the duality within it.

This does not mean that I am saying anyone should look past the story of Jesus Christ—who he was, what he has done, what he believed in, what he taught, and how he existed, both in biological and existential form. Because if we were to do so, we would ignore everything both the light particles and waves have to offer, which are essential for us to survive healthily and prosperously as a species in this æon.

The internal aspect of the wall, where some do not have slits to provide the way, represents those who are unfaithful to the notion that there is a way behind the wall. The external aspect of the wall is the influence of those around them, and albeit frequencies that we bathe in and societal norms built upon envy, pursuit of success, and dominance. Which, when considering a deeper understanding of that, are *truly* just external influences built upon sinful desires and the resulting feelings experienced from them.

My goal is not to convert people. My goal is to describe what is scientifically proven about the æon in which we reside, the psychology and sociology of mankind, which regardless of opinion, cannot be denounced. Many other religions and belief systems will explain and depict Jesus Christ as someone who is confined to only a portion of spirituality and a fraction of what needs to be taught, but this simply is not true.

When we speak about progression as a species, advancement in civilization, and the ability to coexist cooperatively with more intellectual and powerful beings—regardless of biological form—his teachings are essential for our passage to that type of species and civilization. Anger, jealousy, destruction, dishonesty, and deceit, among other sinful acts, will always provide a wall that the great filter uses to halt humanity's progression. Therefore, it is only logical to slice slits through it at a spiritual, societal, and technological level.

Let's start with slicing slits through the spiritual version of the wall. What stops an individual from loving their neighbor, forgiving those who have deceived us, and those who have committed malicious acts against us? What divides us from living sinless lives versus adopting sinful mindsets and actions? Although this can be influenced by external forces, it is internally that we decide which side of the wall we reside on.

Again, this does not mean that I want you to worship and believe in Jesus Christ, pray with or for Him every day, or convert to Christianity, Catholicism, or any other religion that depicts Him as the one and only. It is His teachings that, if everyone abided by—albeit we might face strong retaliation from the wall of the great filter that has been built up—we would eventually succeed in residing on the waveform-side of that wall. Once we are on the waveform side, we can see and choose every single point in time and space that the particle resides in. In this state, we would be limitless, never bound by time itself.

This internal wall, at its simplest and least complex explanation, is built upon evil. We all know what evil is, and we know that evil resides in the sins that religion teaches us to avoid. The issue with the internal wall is breaking it down and removing it. That is why people often say, "Oh, I drank again," "I smoked again," "I ate chips, ice cream, fast food, and sat there again," "I did this sinful or harmful thing to myself yet again," or "I hooked up with a random person again. I'm weak." Most of the time, people know and admit that they are weak. At times, and within the current generation, this is said laughingly to dismiss the natural feeling from the heart and mind of certain actions being unjust and incorrect. Everybody knows many individuals will laugh when hurt to avoid the pain, and emotional pain is no different when the pain is brought on by oneself. "I took my anger out on someone else," "I expressed my internal issues through negative actions," and "I hurt someone because of my own flaws, therefore I am weak." Are examples of individuals that will project their personal issues on others in an attempt to expel what they need to feel. I have experienced this with individuals who randomly blurt out and harass me specifically or personally, when I neither know them nor have done anything wrong. This is both caused by their personal issues being unresolved, possible substances and the withdrawal due to them, and the ability to maintain level-headedness in situations like that. The wall that divides the wave from the particle does a good job at taking advantage of those that are weak, and eventually does not need the person's compliance to exude such malicious and unwarranted behaviors.

However, these individuals are not necessarily always weak. They simply do not comprehend the wall that invisibly stands in front of them, internally blocking their way. Even though many have been shown how to break down that wall, remove it, and maintain its destruction, they struggle to do so. Yes, in some ways, this means they are still weak, but it does not mean they cannot strengthen themselves to break it down eventually.

As someone who chose celibacy for years, who resisted external negative influences, and when those influences occurred, acknowledged them, fought them immediately, and battled against them, I, AJ, can attest to the internal wall that resides within all of us. It continually tries to strengthen itself to block us from the divine light, frequencies, the mindsets of Jesus Christ with all of its teachings, and the Aeon of Pisces.

Once someone breaks down that wall, fights against it, and resists the external influences that affect their internal mindset, the challenge only becomes greater. Breaking down that wall means war—internally and externally. The internal battle, over time, becomes easier, but the external battle is more difficult to control due to technological advancements and the influences of other individuals who use various methods and substances in today's world to prevent you from maintaining the wall's destruction.

The first step is to differentiate between the internal wall and the external influences that create both walls. Once you have broken down that internal wall, it is imperative to break down the external one. To do this, you must remove yourself from the constant bombardment of societal norms and the technology we are entangled in. I don't mean putting on an aluminum foil hat and hiding in the woods. I mean step away for a day, or even just a few hours. Leave it all behind, even for a short time, and test yourself. See what has made you feel a certain way and what you're truly experiencing.

The test is quite simple: distance yourself from all the frequencies, vibrations, and influences that society offers. Though society's aim is often to progress positively, there is also a pull toward negativity, which brings destruction and works against our higher goals. Take this time to contemplate how the external forces impact you, and then, once removed from those influences, observe yourself again.

In certain scenarios, when you are far removed from these frequencies, vibrations, people, and societal influences, you may realize that many aspects of your internal and external walls do not actually come from you. Whether they are societal norms instilled within you, making you feel they are internal, or external forces constantly bombarding you to keep you in a specific mindset, these forces shape the walls around you. It is only when you remove yourself from the world around you, even for a brief period, that you become capable of perceiving and acknowledging these external forces for what they truly are. Once you can recognize them, repeating the experience of stepping away allows you to create space to battle against those forces. Yes, it may feel like war—a difficult, ongoing struggle—but it will ultimately free you when you know for certain what is truly yours and what has been imposed upon you.

When removed, contemplate all scenarios, mindsets, situations, emotions—everything that has made you feel disconnected from yourself, distressed, sad, angry, or opinionated. Sit with these feelings and reflect. Look at these mindsets, opinions, and emotions, and wait to see if you receive a similar response, reaction, or similar information and opinions. This will provide a clue as to which external inputs have been influencing aspects of your life that may not have been prominent or important without them. It will allow you to think for yourself, even if only for an hour or so, offering solace to your mind, body, and soul.

However, don't just contemplate past events, potential obstructions, or external influences. Think about future opinions, mindsets, and plans for your life, the individuals within it, and the world around you. When you return from your journey—whether it be a few hours, days, or weeks—back to the world bathed in frequencies that others can use to influence you, ask yourself the same questions again. They won't know the answers you discovered on your own.

This external wall is more difficult to overcome because it is intertwined with societal norms, success rates in life, and the "luck" that many successful individuals describe. There's a reason for this: those who abide by and submit to these external forces are, at times, rewarded. This reward is shaped by the person's personal goals and desires. It doesn't mean that submitting to the wall makes you a millionaire, but rather, you receive something you've been striving for. This is often tied to emotion, distress, or a lack of essential needs for survival.

I, AJ, was homeless for about three weeks. By then, I had built a bulldozer for this external wall—which inevitably led to my homelessness. For years, I won awards, broke records, and worked hard enough to earn more annually than my higher-ups. This isn't a boastful statement; it was a necessity for me to understand the nature of this wall. Aside from a few global and regional competitions, I was not frequently recognized as the best. However, in the corporate world and private sectors, more often than not, I was. The wall became stronger and more pronounced as I gained recognition. Eventually, certain individuals I had entrusted within my workforce forced my dismissal through their networking and lack of strength. At the time, I was unaware of this, but as the years passed, I was informed of it. This was my personal experience with the external wall, which goes beyond just residing in frequencies.

Once I realized that individuals could use technology to influence and harass myself and others, I spent years studying how it was done. Many people throughout my journey provided clues, both inadvertently and deliberately, on how this occurs and how to combat it. This is partly why I began writing my book, although my deployment with the U.S. military also inspired me to write down my advice for others to read and, hopefully, implement.

Beyond direct external input, there is also indirect external input. We now know that, through pulse-modulated microwave frequencies, anyone with 10 hours of knowledge and some basic, low-cost equipment can harass anyone on Earth. And that was with 1970s technology. In today's world, using AI and the internet, someone can implement these methods instantly and effortlessly. This is where the more complex challenge arose for me, though not for long. Given everything that's been discussed about frequencies, information encoding, and methods like V2K, I don't need to elaborate further on other external forces at play.

Leaving the "bath" of frequencies and technology is crucial as a stepping stone to overcoming the entire wall that separates us from the wavefunction of Pisces. Faraday materials can help block certain frequencies, though they aren't absolute in uncovering all external forces at work. However, tools like these can aid the strong-minded in becoming aware.

This external wall is more difficult to overcome because it is intertwined with societal norms, success rates in life, and the "luck" that many successful individuals describe. There's a reason for this: those who abide by and submit to these external forces are, at times, rewarded. This reward is shaped by the person's personal goals and desires. It doesn't mean that submitting to the wall makes you a millionaire, but rather, you receive something you've been striving for. This is often tied to emotion, distress, or a lack of essential needs for survival.

Nature's Hard Drives and Remote Downloads of Information

The idea that light carries information is not just theoretical—it is a reality that has been proven through both scientific and mystical means. In the modern world, light is used to transmit data through fiber optics, Li-Fi (Light Fidelity), and other technologies that encode information into light waves. But the concept of light as a carrier of information goes back much further than the advent of modern technology. Throughout human history, mystics, philosophers, and scientists have recognized the power of light to convey knowledge and truth. In ancient Egypt, for example, light was associated with the god Ra, the creator and sustainer of life. The Great Pyramids, built with stones rich in quartz, were believed to harness and focus light, using it as a tool for spiritual and physical transformation. Today, it is understood that light can be modulated in various ways to encode data. This process involves altering specific aspects of the light wave, such as its amplitude, frequency, or phase, to represent different pieces of information. The same principles that allow data through fiber optics to be sent also apply to the natural world, where light from distant stars carries encoded information about the universe.

"Crystals are living beings at the beginning of creation"
- Nikola Tesla

Considering the more recent discovery of altermagnetism, it is fully accepted and understood that it is quite feasible for light to encode data and information within quartz crystals, which can be retrieved at a later date. However, quartz is not the only crystal capable of being encoded with such data. Individuals like mystics, psychics, astrologers, and spiritual healers, who typically lean toward this side of the now-scientific spectrum, cannot be disregarded by the scientific community as a whole. Humans emit photons, low-voltage energy, and slight magnetism from their bodies and biofields. A person's strong emotions, willpower, mentality, and knowledge can encode or program a crystal based on their desires, emotions, and thoughts.

Altermagnetism is still in its infancy, but that does not mean individuals with the right strength, power, and knowledge cannot encode various types of crystals with the information they wish to imprint. It also implies that light, whether from the cosmos, Earth, or elsewhere, can encode a crystal, quartz, or other materials with altermagnetic properties. Just because humanity currently lacks the technology to read, retrieve, and translate that information into something comprehensible does not mean it isn't there.

This is why certain individuals, particularly those who are empathetic, possess high energy levels, or strong emotions, tend to gravitate toward crystals, whether birthstone crystals or clear quartz. They often speak about the powers these crystals emit into the world around us. While not every mystic, psychic, or spiritual person will be truthful in this regard, those with the proper demeanor and palpable energies tend to be more credible. Ultimately, the concept of programming a crystal through the human body can no longer be ignored.

"In crystal we have a pure evidence of the existence of a formative life principle, and although in spite of everything, we cannot understand the life of crystals - it is still a living being."
- Nikola Tesla

The Spiritual Significance of Light

In many spiritual traditions, light is seen as a symbol of divine wisdom and enlightenment. The phrase "light of the world" is used in Christianity to describe Jesus Christ, who is believed to bring spiritual illumination to humanity. Similarly, in Hinduism, light is associated with knowledge and the dispelling of ignorance, represented by the god Krishna, who is often depicted with a radiant aura.

The connection between light and spirituality is not just symbolic—it is also experiential. Those who meditate or engage in other spiritual practices often report experiences of inner light, a radiant energy that fills the mind and body with a sense of peace and clarity. This inner light is thought to be a reflection of the divine, a glimpse of the higher consciousness that pervades the universe. Light is also seen as a bridge between the physical and spiritual worlds. In mystical traditions, it is believed that light can carry messages from higher realms, providing guidance and insight to those who are attuned to its frequencies.

Whether through visions, dreams, or other forms of spiritual communication, light serves as a medium through which the divine can speak to us.

Notes

Chapter 7: The Divine Medium and the Evolution of Cosmic Communication

Before we speak about specific influential figures throughout humanity's existence, we will discuss artwork provided through the centuries that depict what we have discussed so far. Artists from centuries before this day had limited knowledge and understanding of technology and the world around us, yet they still were able to depict certain things that are explained by both my theory and current technological advancements and knowledge.

The Medium as a Divine Conduit

The universe is not just a vast expanse of empty space; it is a medium through which information, light, and energy travel. This medium connects all things, from the smallest particles to the largest galaxies, creating a network of communication that spans the cosmos. In this medium, light is more than just a source of illumination; it is a carrier of information, capable of encoding and transmitting messages across vast distances.

The connection between the universe and the human brain is particularly striking. When looking at a current model of the known universe, with its vast networks of galaxies connected by filaments of dark matter and light, we see a pattern that closely resembles the structure of the human brain. Neurons and Neurites in the brain are connected by synapses, through which electrical impulses (information) travel, creating thoughts, memories, and consciousness. Similarly, galaxies are connected by light and gravitational forces, creating a cosmic consciousness that transcends time and space.

This concept challenges our traditional understanding of communication. Light, which travels at an astonishing 299,792 kilometers per second, is still considered slow in the context of the universe's vastness. The stars and galaxies observed in the night sky are thousands, sometimes millions of light-years away. Meaning much of the light seen may have left its source long before planet Earth developed into what it is known as today. Yet, there must be a way for information to travel faster than light, allowing for instantaneous communication across the cosmos.

In this context, considering how long it takes light to travel to Earth, we must explore a quicker and more immediate way in which God, the divine, and other cosmic intelligences communicate across such extreme distances. Given that most stars and galaxies are thousands to tens of thousands of light-years away, it seems unlikely that beings of higher power would rely solely on light to communicate. There must be methods to send information instantaneously, surpassing the speed of light.

One possible explanation involves the idea that once light reaches an object, it creates a channel and pathway to that object. Photons, which do not experience time, could potentially deliver information along the light's wavelength without delay. As photons travel along a path previously established by light reaching a distant cosmic object, they could deliver messages or data instantaneously, bypassing the need to wait for time to pass. The idea is that the photon sent along the same path as its predecessors could exploit this pre-established pathway to transmit information instantaneously.

The Theory of Rifled Light

Considering the theory that light can transmit information once the wave and particle have reached an object or destination, one cannot dismiss the idea that the object or being is attuned to absorb that information accordingly. For example, when UV light reaches human skin cells, our DNA is coded to produce Vitamin D. UV light is just one of many forms of light that reaches Earth everyday from the cosmos. Some forms of light flow through us, some forms of light bounce off us and other objects, consequently forming part of the medium that is unseen and surrounds us all. How does humanity know if DNA is encoded with the ability to absorb and respond to certain frequencies or wavelengths of light that provide our bodies and souls with information, potentially creating and causing various forms of divinity and abnormalities? I am certain that humans have the coding somewhere within DNA, ready to adjust accordingly based on the absorption of specific particles, frequencies, and wavelengths. But waiting many years, decades, or even centuries for a message to trigger a reaction within our bodies seems rather archaic in some ways. The key to progressing and understanding both divinity and how the cosmos has shaped humanity and sentient life throughout history lies in the transmission of information through light and frequencies in ways that are faster and more advanced than what is currently perceived.

Consider this metaphor using technological progression in weaponry. The cannonball, for example, is shot out of a cannon with minimal spin, relying on brute force to reach its target. It's a simple object, with its significance being its inertia overcoming gravity and the drag of the air around us. The explosive propulsion pushes it toward its destination. However, once humans understood how to make an object move through the air with accuracy and speed, without relying on great mass, the bullet was born. The barrel of a gun has what is called rifling—grooves within the barrel that provide spin to the bullet. This spin allows the bullet to cut through the air and reach its destination with accuracy and speed. Similarly, progressing to the instantaneous transmission of information through light will follow a similar evolution. The "rifling" of the waveform of light, providing the particle with information at nearly instantaneous speed, is our next step. Even with the most up-to-date theories and physics on information transmission through light and frequencies, Earth still lives in a world where humanity uses the cannonball to send information.

The rifling of our information will undoubtedly involve a combination of quantum mechanics and the use of the medium or quantum field, not just the modulation of frequencies. "Hacking" or "tapping" into the way photons perceive time, which is a lack thereof, is key to producing our first rifled barrel of information. Gravitational lensing can provide an aspect of rifling, but the true "barrel of the gun" is at the origin of the photon's journey. It is the guide for the pathway that the photon travels upon, even if the photon traverses a medium without deviating influences. The barrel is the waveform, but what is the rifling? If you consider how electrons are seen, and how they react with the medium or quantum field, they are the first part of how to rifle information. Even though an image of an atom is depicted in a similar way as planets orbiting a star, this depiction is very misleading. Electrons perform quantum teleportation, and are technically in superpositions before being observed where they are. That is why they jump around an atom in various positions, sometimes even swapping with other atoms and jumping to other areas of the medium or quantum field. These electrons hold information, such as spin. The way they perceive time, and traverse the medium, is the way information is rifled to its destination. Entangled particles with specific states that can be expressed between each other at vast distances instantaneously is when humanity progresses from the cannonball to the bullet.

The important factors in rifling information through methods that the electron uses would be to know how and where the electron or particle will be observed. If one knows this before it is sent, one can encode information to the electron or particle that dictates a result once it arrives at its destination. The information encoded in it will change the function of the object once it reaches its destination and is absorbed. If the absorbing object is only attuned to, for example, a spin downward, and throughout the wave function or beam of light there is only one destination or object that is attuned to a downward spin, then it should be capable of jumping to that object instantaneously without the need to travel through space or for time to pass. Now consider an article of light, or photon. Within itself, it does not have a clock and does not abide by the imprisonment of time. Another way to "rifle" light is to focus it using a laser beam at close proximity.

The Cosmic Connection in Religious Art and Texts

One aspect of my thoughts on my own Theory of Universal Divine Light and Frequencies is: Considering the biblical reference for dividing the people into multiple languages, multiple regions, multiple ethnicities, and multiple beliefs, one must ponder the thought that light was refracted due to gravitational lensing and reaches Earth in multiple paths initially. These multiple paths would consider the division between humanity and its influence on the species as a whole.

The angels or UAPs that visit Earth as messengers are meant to realign the spread of information. Individuals like Jesus, and the focal aspects of gravitational lensing in each precessional age, provide insight into each aspect of what the true and intended purpose of the light's message was.

Even though gravitational lensing can scatter information, depending on certain areas of the Earth and specific celestial alignments that modulate light and information with their own gravity, it is easy to comprehend how specific areas of the Earth could receive focused light and information instead of scattered light and information. This raises the question: Did ancient civilizations worship celestial alignments at specific times throughout the year for a good reason, and possibly even a scientifically proven one? It seems so. If you look at the illustrations from Chapter 3, you'll see how many celestial bodies, though not perfectly aligned, are close to the Sun as it passes in front of Pisces. If we are to assume that a more advanced intellectual being or civilization in the cosmos needed to send information or updates to those already here, to have a biological effect on specific individuals or humanity as a whole, they would utilize astronomical alignments during a time on Earth where electromagnetic fields could help carry that information properly. That is most likely why so many different cultures share a similar story. The story is written in the stars—a place that can't be burned or erased by the next regime, civilization, king, dictator, or terrorist organization. It's not a book that can be burned; it's a book that is burned into our sky, through the cosmos, forever being edited on its way to us, most likely instantaneously.

This could explain the rise and fall of civilizations, the jumps in technological advancements, the leaps in intellect, and the origins of divine intervention. To prove this theory, one must perform an experiment where a similar thing can occur. However, if you view all of the technological advancements that are used daily, there may not be proof on a macro scale like the cosmos, but there is current proof on a micro scale compared to the cosmos. The entire population carries something in their pockets every day that uses similar technology. Radio waves and cell towers send signals at the speed of light through the air, invisibly, carrying extreme amounts of data—from videos to photos, literature, and live, low-latency video calls. And we're not even using lasers or any form of light to send this information. In religious literature, across many different religions, angels are sent down here to guide and watch over humanity. They don't make decisions for humanity; they guide us. They don't necessarily edit our genes or intervene with our DNA, but they guide us.

At the same time, guardian angels are entities that were all created at one time, all assigned to each of us. They make a decision upon their creation of whether or not they will protect and guide the human being to whom they are assigned. They make an instantaneous decision—yes or no. If it's a no, they are dismissed, and the human being gets a different guardian angel. If they say yes, they wait, sometimes for millennia, for the human being to be born.

In Roman Catholicism, as elucidated earlier, angels are beings of pure light, and guardian angels are essentially no different. They accompany you from well before your birth until the day you die, guiding you through life according to the blueprint encoded in your DNA, which dictates the decisions you are likely to make. This does not negate the existence of free will, but rather, these angels are meant to guide and shield you from negative influences and the potential poor decisions that could arise from your genetic makeup, your DNA composition, and the inherent traits that shape your personality. Alterations to one's DNA or genetic structure can potentially disrupt this intrinsic alignment with one's guardian angel, thereby altering the coherence between oneself and their celestial guide. This is a concept that the Roman Catholic Church has advocated for quite some time. They are not entirely mistaken. However, the traditional depictions of angels as feathered, winged beings with human-like features soaring through the sky are, to some extent, inaccurate.

It is important to recognize that the universe functions as a neural network on a macro scale. Our brains, while still more powerful than most computers, particularly in areas such as creative thinking and emotional reasoning—functions that go beyond mere arithmetic or AI-generated art—are difficult, if not impossible, to surpass. Supercomputers may be approaching this capability, but they are not yet on par with the human mind. The universe, however, represents a cosmic, macro-scale neural network so vast that it extends to the farthest reaches of the observable cosmos. This universal intelligence could be considered the one entity capable of surpassing the computational power of the human brain, as well as the cognitive faculties of any other sentient being.

God is often depicted as a white-bearded man, but even if He has manifested in such a form, He transcends this image. People will come to understand that even the angels that traverse our skies—those that evade capture by our military and are now openly discussed—are themselves guided by this divine force. They are its messengers, and whether visible or not, these angels, who interact with everyone daily, do not possess ultimate control. In all systems of life, there exists a hierarchy, and this extends to intellectual beings as well.

Thus, if an intellectual entity from a distant star or across the universe wishes to communicate with us, it need not wait for light to reach us. The only necessity for waiting for light to reach Earth is to establish a direct, lag-less conduit, a pathway for particles encoded with information, therefore enabling instantaneous data transfer. God is the Alpha and the Omega, the beginning and the end. This statement can be understood as referring to the first speck of stardust—the primordial molecule of the known universe. When the universe eventually collapses upon itself, this speck will also be the last. In the same way that Jesus Christ represents the Alpha and Omega within our world during this age, God should be envisioned as the initial molecule, which, from the dawn of time, possessed a mass equivalent to the entire known universe, and will ultimately be the last as the universe contracts back into a singularity.

Jesus Christ is the Alpha and Omega of his era here, and regardless of differing beliefs, we have many more years to spend in his era. Most other religions originated, evolved, and thrived during different ages. It is not within my power, your power, or anyone else's—whether they read this book or not—to alter that.

For centuries, humanity has depicted divine beings, angels, and gods through art and literature, often drawing connections between these figures and celestial phenomena. From the Vedic texts of ancient India to Renaissance paintings, there is a recurring theme of celestial beings interacting with humanity. These depictions raise questions about the true nature of these beings—are they merely figments of human imagination, or do they represent encounters with divine beings that watch over the population?

In ancient Hindu and Indian religious texts, Shiva is portrayed residing in a flying chariot, known as a Vimana. Vimanas are clerically depicted through written language as flying chariots or celestial vehicles with remarkable abilities, representing the divine transport used by gods and other mythological beings. The concept of Vimanas dates back to some of the earliest Hindu scriptures, where these flying machines appear in various forms. Over time, their descriptions evolved, becoming more elaborate and technologically advanced in later texts. Over time, the translation and interpretation of language changed. Scholars and individuals in power migrated from actual flying chariots to the tops of Hindu temples, which resembled a Vimana. Hindu temples were used to allow both living and deceased individuals to transcend into the heavens and traverse the medium at will. It was used as a conduit for the soul to enter the afterlife, and became a sculpture of significant religious events instead of a chariot in the sky.

Vimanas first appeared in ancient Sanskrit literature, specifically in the Rigveda, one of the oldest known texts of Hinduism, dating back more than 3,000 years. In the Rigveda, flying chariots are mentioned as the vehicles of gods, used to travel across the heavens. These celestial chariots allowed the deities to traverse the skies effortlessly, engaging in battles or visiting the earthly realm.

Later texts, such as the Ramayana and the Mahabharata, provide more detailed descriptions of Vimanas. In the Ramayana, one of the central stories involves the demon king Ravana abducting Sita, the wife of Lord Rama, using a flying machine known as the Pushpaka Vimana. This flying chariot, which was originally owned by the god Kubera, was said to be golden in color and possessed the ability to fly through the air and travel at incredible speeds. The Pushpaka Vimana was described as not only swift but spacious enough to accommodate many people. It became a symbol of divine or royal power, a vehicle that could defy the normal constraints of earthly movement.

The Mahabharata, another major epic of Hindu literature, also references Vimanas. In this text, the flying chariots are frequently depicted as divine vehicles used by gods and heroes during great battles. For example, Arjuna, the Pandava prince, is known to have been taken into the skies by a Vimana called a "celestial car," which allowed him to witness vast landscapes and engage in combat from the air. The Mahabharata describes such vehicles as capable of moving freely in any direction, showcasing remarkable speed, agility, and combat capabilities. In some interpretations, these Vimanas were said to be able to traverse not only the Earth's skies but also other worlds, giving them a cosmological significance far beyond simple transport.

The descriptions of Vimanas in ancient texts vary, but they often emphasize their grandeur and technical capabilities. Pre temple interpretations of the texts suggest that these flying machines were not merely mythological symbols but were portrayed with mechanical details, giving rise to speculation that ancient cultures might have envisioned—or even known about—technological advancements far ahead of their time.

In the Samarangana Sutradhara, an 11th-century Sanskrit text on architecture, there are passages that describe Vimanas in a technical manner. This text details how Vimanas are constructed with mechanical components and powered by engines fueled by an unknown energy source. The descriptions include not only the ability to fly but also the capacity for these machines to traverse great distances quickly, carry weapons, and operate without the need for external pilots. There are even references to their ability to be stored when not in use, implying a sophisticated system for managing these flying machines.

The Pushpaka Vimana from the Ramayana is described as a golden, palace-like chariot with a large interior, capable of transporting many individuals. It had the ability to fly wherever its user commanded, underscoring the advanced nature of this vehicle. The Vimana was not simply a flying machine; it was also an emblem of royal or divine power. Interestingly, while the Vimanas described in both the Ramayana and Mahabharata possess technological qualities, their usage and control are often tied to individuals of high spiritual or royal status, emphasizing the connection between divine favor and access to this advanced technology.

In modern times, these descriptions have sparked considerable interest and debate. Some theorists suggest that Vimanas could be ancient references to advanced technology, possibly scriptures depicting sightings of now well known UAPs. These interpretations are often based on the detailed mechanical and operational descriptions found in texts like the Samarangana Sutradhara. The text speaks of these machines as having a complex design, powered by what is translated directly as fuel or energy sources unknown to the world.

Scholars, however, generally view Vimanas as symbolic rather than technological. They argue that these flying chariots are metaphors for divine power and mobility, representing the gods' ability to move between realms, including the earthly, celestial, and spiritual planes. In this view, the Vimanas are closely tied to the religious and cultural context of Hinduism and are not meant to be interpreted as literal technological devices.

While the scholarly consensus remains cautious about such interpretations, the vivid depictions of Vimanas in ancient texts like the Ramayana and Mahabharata provide a tantalizing glimpse into how early cultures may have seen the flight of the divine during their cosmic travels. Whether seen as mythological symbols, divine beings, or possible representations of ancient technology, Vimanas remain an intriguing element of Hindu tradition, inspiring both awe and curiosity across the ages.

None of You Read The Bible, Don't Lie! For Shame!

If you review gospels in the bible, specifically The Book of Revelation, there are many descriptions of Angels and Beasts that depict examples of divine light influencing mankind. However, it is through art with which humanity today can find correlations between what humanity has observed and photographed over the past century, and what ancient civilizations illustrated and described through written language and artistic skills in their pre-industrial, pre-informational, pre-modern and technological time.

In Catholic art we see angels descending from the heavens, often depicted as beings of light, auras, and mechanical objects in forms that could be interpreted as UAPs. For instance, the famous painting "Madonna with Saint Giovannino" from the Palazzo Vecchio depicts what appears to be a UAP in the sky, with a man and his dog gazing at it. Art historians and religious scholars have long debated these images, often dismissing them as symbolic representations. However, considering our current understanding of light, frequencies, and the medium in which they travel, it becomes increasingly plausible that these depictions were based on real encounters of divine intervention. The genetics of ancient Indians and Hindus in comparison to Mediterraneans and Catholics expresses how they were interpreted by different genomes throughout multiple generations. Each culture and ethnicity viewing similar phenomena through their own abilities, expressed their observances through different artistic and literary talents. Mediterraneans and Catholics typically depicted what they saw through ultra realistic art and illustrations (according to their era), while Indians and Hindus explained their encounters with divine entities and deities through literature. Both are significant and intellectual, yet artistic depictions allow us to speculate less about what was seen and what was interpreted by their ancestors. Without transparency from modern governments about UAPs, this argument would fall within the realm of conspiracy theories. However, since the USA now openly discusses this with their Congress, among other nations, it cannot be dismissed or ignored so easily. This does not denounce divinity, God, or anything of the sort. It is only meant to describe how different cultures, genomes, and generations with different talents, technological advancements, and beliefs depicted, interpreted, and portrayed their observances. This is why physics, science, and the mathematics supporting all subcategories are so important. Previous iterations of a diverse humanity, without a common language, immediate and precise communication, and collaborative research, most likely described similar instances of divine intervention in vastly different ways that were attuned specifically to their cultures. Without mathematics and knowledge of how light and frequencies can be encoded with information, previous iterations of humans could only rely on experiences in their lives and the understanding of the world around them described by their era. However, illustrations through art and early photographs were difficult to alter. Text and written language, when translated, can be easily deviated from its original meaning. Artists that paint realistic images based on text and observance typically express their ideas based on personal experiences and common knowledge of their own society, through overall techniques of their age.

"The Baptism of Christ"
by Aert de Gelder, 1710

Scholars generally interpret "The Baptism of Christ" by Aert de Gelder (1710) within a traditional religious context. The disc-like object in the sky, often identified by UAP enthusiasts as a possible depiction of a flying saucer, is typically understood by art historians and theologians as a representation of the Holy Spirit or Angels descending from the heavens to witness and support the baptism of Jesus Christ.

In Christian iconography, the Holy Spirit is often symbolized by light or a dove descending from the heavens during significant religious moments, such as the baptism of Jesus by John the Baptist. In de Gelder's painting, the glowing disc emitting rays of light is considered to be an artistic representation of this divine presence, consistent with the religious symbolism of the time. Scholars suggest that the rays of light emanating from the object are meant to illustrate the descent of the Holy Spirit unto Christ, marking his divine anointment during the baptism. As we have discussed, scientifically speaking—according to modern understanding—this can also be interpreted as information that encodes a biological form to adhere to specific traits that our DNA is innately designed to receive. God himself would know how to communicate and provide divine effects to our DNA through frequencies and light, similar to how information is sent through lasers from one satellite to another to transmit information and data. The only difference is humans do not know, openly, how to effect a biological form to trigger DNA alterations with such methods. If it is or eventually is discovered, those with the knowledge should *never* tell anyone. *Never.*

"The Madonna with Saint Giovannino"
Domenico Ghirlandaio or his school, 15th century

This is a painting of the Madonna with Saint Giovannino and the child of Jesus Christ. However there is an obvious disc in the background of the image, with a man and his dog viewing the supernatural phenomena. Scholars explain this as a representation of an angel or divine messenger. This interpretation fits within the broader Christian iconographic tradition where divine figures, such as angels, were often depicted in symbolic or abstract ways. The glowing, disc-like object in the background could be seen as a more ethereal or symbolic representation of an angel descending from heaven, not necessarily shown in human form, but instead as a heavenly presence. God has his messengers, and releases them in various forms selected in many ways throughout humanity's existence.

The man in the background looking up at the object, along with his dog, represents a witness to this divine event or apparition, and an acknowledgment of the supernatural. In religious art, angels and other heavenly beings were sometimes represented with non-literal imagery, such as clouds, light beams, or abstract forms correlated with many attributes of living beings to signify their otherworldly nature. For ancient humans did not have mechanical devices and machinery to compare these observations to—only living beings.

"The Annunciation"
By Carlo Crivelli, 1486

The most notable feature in Crivelli's "The Annunciation" is the beam of light that emanates from a circular object in the sky, aimed directly at the Virgin Mary. This circular object is typically interpreted by scholars as a symbol of the Holy Spirit, and the light beam represents the divine intervention through which Mary conceived Jesus.

The dove at the center of the beam, representing the Holy Spirit, supports this interpretation. In Christian art, the dove is a frequent symbol of purity and the presence of God, tying directly into the theme of the Annunciation.

The painting is imbued with Christian symbolism, much of which points to theological interpretations of the Virgin Mary's role as the Mother of God. The circular object (interpreted as a celestial realm or a symbol of God's presence) and the direct beam of light are both typical of religious artworks of the time that aimed to visually convey spiritual mysteries.

This depicts a correlation with the methods with which the divine communicate with humanity. Through concentrated, focused, and coherent light or frequency transmission, the Virgin Mary could have easily experienced a form of information transmission that triggered a biological reaction. This biological reaction to light or frequencies may have allowed God to choreograph Jesus' conception with his image and, in turn, Jesus' divine characteristics. Using his Angels that descended upon earth, God had his own conduits of divine intervention at play.

Notes

Chapter 8: Quantum Communication: Because Yelling at the Universe Takes Too Long

The vastness of the universe presents significant challenges when it comes to the transmission of information across interstellar distances. Light, the fastest way to traverse the medium that science knows, still takes years, decades, or even millennia to travel between stars. Yet, in the pursuit of understanding and perhaps even communicating with distant civilizations or celestial phenomena, we must explore the possibilities of transmitting information at speeds that outpace the traditional constraints of light-speed travel. Quantum Entanglement expresses a way that the universe communicates through the medium while ignoring gravity, space, and time.

Quantum entanglement is a confirmed phenomenon that occurs when two particles become connected, and the state of one particle instantly influences the state of the other regardless of the distance between them. These particles can communicate with each other faster than the speed of light—though it's not about actual communication but rather an intrinsic connection between their states. This "spooky action at a distance," as Einstein called it, suggests that information can be transmitted through the medium, challenging our and his understanding of gravity, space, time, and the medium all around us. Entanglement provides an important clue to the mechanism for transmitting information across vast distances, connecting us to the stars and beyond.

Einstein famously called it "spooky action at a distance" because it seemed to violate his theory of relativity, which says that nothing can travel faster than light—not even information. However, entanglement doesn't transfer information in the traditional sense, so it doesn't break the laws of general relativity. Let's take the example of two particles, say electrons, which can be entangled. Each of these particles has a property called "spin," which is similar to having its own microscopic magnetic field that can point in different directions—usually labeled as "up" or "down." Normally, when you measure a particle's spin, it could either be up or down based on probability which is a fundamental principle in quantum mechanics. But when two particles are entangled, their spins are mysteriously connected. If you measure the spin of one entangled particle and find it's "up," the spin of the other particle will instantly be "down," regardless of whether that particle is next to you or on the other side of the galaxy. Before you measure them, both particles exist in what is called a superposition—they're both spin-up and spin-down at the same time. It's only when you measure one particle that its entangled partner's state becomes determined.

This phenomenon was first proposed by Einstein, Podolsky, and Rosen (formally called the EPR paradox) as a thought experiment to challenge the weirdness of quantum mechanics. They argued that quantum theory must be incomplete because it allowed for this "spooky action." However, in the 1960s, physicist John Bell formulated Bell's Theorem, which provided a way to test if the predictions of quantum mechanics (including quantum entanglement) were correct. This led to experiments that overwhelmingly confirmed entanglement's existence, denouncing Einstein's skepticism.

When two particles become entangled, they share a quantum state. This state contains all the information about their properties. Even when the particles are separated, the shared quantum state remains intact until one particle is measured. Upon measurement, the shared state "collapses" and both particles take on definite values that reflect their entanglement.

This collapse happens instantaneously, no matter how far apart the particles are. It's as if the universe doesn't acknowledge the distance and time separating them. Quantum mechanics transcends space and time, in similar ways that those who traverse the medium do. While the particles are instantly connected, you can't use this phenomenon to send information faster than light. The correlation between entangled particles doesn't transmit usable data on its own, but expresses the ability to traverse the medium and utilize faster than light travel.

Quantum entanglement cannot be explained by traditional, classical physics. It's one of the most baffling aspects of the quantum world that has been repeatedly verified through experiments. Entanglement is crucial to our understanding of the quantum world. It's not just a weird glitch in the matrix; it's a core feature of how particles behave at the smallest scales. Entanglement is the key to quantum computing. In a classical computer, information is stored in bits that are either 0 or 1. But in a quantum computer, qubits can be 0, 1, or both at the same time. This is called "superposition". When qubits are entangled, operations on one qubit can instantly affect others, allowing quantum computers to perform complex calculations at incredible speeds. Entanglement is also used in quantum cryptography. In quantum key distribution (QKD), entangled particles ensure that any attempt to intercept a message will be instantly detected, making eavesdropping impossible.

Quantum entanglement challenges our understanding of reality at a fundamental level. How can two particles be connected across space without anything physically linking them? In classical terms, it makes no sense. But quantum mechanics doesn't follow classical rules—it works in probabilities, wave functions, and states that defy common sense.

The phenomenon has made many physicists consider that the universe may be far more interconnected than initially realized. Some even hypothesize that quantum entanglement hints at a deeper level of reality where space and time might not behave the way it is perceived. In short, quantum entanglement is one of the most counterintuitive, bizarre, and exciting phenomena in the quantum world—and we're just scratching the surface of what it could mean for the future of science and technology.

The implications of quantum entanglement are indeed science-altering, particularly when it comes to our understanding of consciousness and the nature of reality. Since light is a carrier of information, and if quantum entanglement allows some forms of information to be transmitted instantaneously, then it opens up new possibilities for communication and connection that transcend the physical limits of the universe. It also expresses one of the many ways that the universe is pre-wired to communicate instantaneously over vast distances.

Theoretical Foundations: Quantum Communication and Photons

At the core of this chapter lies the theory that photons, the fundamental particles of light, can be encoded with information and transmitted across the cosmos, retaining their integrity and quantum state despite the vast distances involved. Photons are unique in that they do not experience time as we do. From the moment a photon is emitted, whether from a star or a quantum device, to the moment it reaches its destination, it remains in a superposition—a state where it has not yet "experienced" time. This property allows one to theorize that photons could carry information in a way that transcends the usual barriers of time and space. Coupled with quantum entanglement, this idea is not as far-fetched as one may perceive it to be.

Quantum Repeaters and Coherence Maintenance

One of the primary challenges in quantum communication is maintaining the coherence of quantum states over long distances. Quantum decoherence occurs when a quantum system loses its quantum properties due to interaction with its environment, which is particularly challenging over vast interstellar distances. To combat this, we can propose the use of quantum repeaters—devices that extend the range of quantum communication by amplifying and correcting quantum signals without destroying the information they carry. These repeaters would be strategically placed in space, perhaps leveraging naturally occurring phenomena like gravitational lensing to assist in focusing and directing the photons along their intended path.

Utilizing the Interstellar Medium and Gravitational Lensing

The interstellar medium (ISM), consisting of gas, dust, matter, and dark matter, pervades the space between stars. While thinly spread, it offers a potential medium through which quantum information is protected and guided. If the ISM can be mapped and understood in detail, humanity could use its natural properties to channel photons, preventing them from scattering and losing coherence.

Moreover, gravitational lensing—where light is bent and focused by the gravitational field of a massive object, such as a star or black hole—could be employed to further enhance the precision and reach of our quantum communication. By aiming photons so that they are gravitationally lensed by a distant star or galaxy, light can be focused onto a specific target, potentially allowing encoded information to be sent across vast distances with minimal loss of signal integrity.

Encoding Information in Photons

Photons can be encoded with quantum information using various methods, such as polarization or time-bin encoding. Time-bin encoding, in particular, allows for the creation of distinct time intervals (bins) that represent different quantum states. By carefully encoding photons in this way, it can be ensured that they carry the desired information across interstellar distances. The encoded photon, once it reaches its target—whether a star system, a planet, or a biological system—can then trigger a specific, predetermined response. This brings up how light sent from a distant star or planet can be encoded to specifically alter itself and the object it reaches, thus transmitting information and data.

Triggering a Biological Response

There is evidence that certain wavelengths of light can induce positive genetic changes. Consider the example of UV light and its effect on human skin cells, triggering the production of Vitamin D. This is a natural biological response to a very specific wavelength of light. Theoretically, if photons are encoded with information that is understood by a biological system at its destination, it could trigger a similar, yet more controlled, response. This would involve not just encoding information, but also ensuring that the photon interacts with the target system in a way that results in a classical measurement—essentially a detectable, meaningful reaction. For example, if humanity wanted to send a signal to a distant civilization, one could encode a photon with a message that, upon arrival, interacts with the local environment or biological systems to trigger a response. This could be as simple as a signal that generates a specific chemical reaction or as complex as altering genetic expression in a targeted organism.

This suggests that encoded photons could be designed to influence genetic expression in targeted ways, potentially guiding the evolution and adaptation of life forms at their destination. This concept extends to critical periods such as conception and pregnancy, where the influence of light on DNA could shape the genetic development of an organism, leading to specific, desired outcomes.

Combating Decoherence

Decoherence is the process by which a quantum system loses its quantum properties, typically due to interactions with its environment. In the context of interstellar communication, decoherence poses a significant threat to the integrity of the information being transmitted. To combat this, science must explore various methods of maintaining coherence over vast distances.

One approach involves surrounding the photon with other particles or photons that inhibit decoherence, effectively creating a protective shell. Another method might involve utilizing the wave function of the photons, ensuring that they travel along the same path as previous photons, thereby reinforcing their quantum state.

Quantum repeaters, as mentioned earlier, also play a crucial role in this process, as they can amplify and correct quantum signals, preventing decoherence from compromising the information being sent.

Photonic Crystals and Natural Phenomena

Photonic crystals, which are structures that can manipulate the flow of light, offer another potential method for maintaining coherence. These crystals can be engineered to control the wavelengths of light passing through them, thereby preserving the quantum state of the photons. While photonic crystals are primarily created in laboratory settings, similar phenomena might occur naturally in the cosmos, perhaps in the form of crystalline structures within the ISM or on the surfaces of distant planets.

Encoding Photons for Cosmic Communication

The core of this theory lies in our ability to encode photons with specific information. Unlike classical forms of data transmission, where information degrades over distance, quantum encoding allows for the preservation of data across immense distances. By utilizing time-bin encoding, photons can be programmed to transition from a superposition to a specific classical state just before reaching their target. This ensures that the information carried by the photon remains intact until the precise moment it is needed, effectively combating the decoherence that typically plagues long-distance quantum communication.

Maintaining Coherence Across Vast Distances

To further protect against decoherence, quantum computers can be employed to maintain a coherent quantum state throughout the photon's journey. These advanced systems continuously monitor and adjust the quantum state, correcting any deviations that may occur as the photon traverses the interstellar medium. Additionally, the strategic use of quantum repeaters—devices that extend the distance over which quantum information can be transmitted—ensures that the photon remains in a coherent state as it travels, maintaining the integrity of the encoded message.

Gravitational Lensing and Celestial Navigation

Gravitational lensing plays a dual role in this theory. While it can scatter light, it can also focus it, creating natural pathways that guide the photon on its journey. By aligning the transmission with specific celestial bodies, we can leverage their gravitational influence to direct the photon to its intended destination. This method not only aids in the precise delivery of information but also helps stabilize the photon's quantum state, reducing the risk of decoherence.

Navigating Celestial Alignments

The alignment of celestial objects plays a crucial role in the success of this cosmic communication. By timing the transmission of photons to coincide with specific alignments, we can use the gravitational wells created by these objects to focus and guide the photons, ensuring they reach their destination with minimal scattering. This method also enhances the likelihood of maintaining quantum coherence, as the gravitational conditions stabilize the photon's quantum state throughout its journey.

A Grand Theory of Quantum Cosmic Communication

Bringing together the intricate concepts of quantum mechanics, gravitational lensing, and the manipulation of the interstellar medium, I, A.J. DelVecchio, proposes a grand theory of cosmic communication that transcends the limitations of traditional light-speed travel. By encoding photons with information and employing a combination of advanced techniques such as quantum repeaters, time-bin encoding, and the use of quantum computers, it is theoretically possible to transmit information across the vast expanse of the cosmos. This method could enable the ability to reach distant civilizations or biological systems with messages carried through light, delivered with precision and clarity.

This theory hinges on our ability to maintain quantum coherence over vast distances, leveraging both natural and engineered phenomena to protect the integrity of the information being sent. As we continue to explore the mysteries of quantum mechanics and the cosmos, this grand theory represents a bold step toward understanding how humanity might one day communicate with other intelligent beings across the universe.

Notes

Chapter 9: Divine Darkness Gravitates Around All of Us

Dark matter and dark energy, particularly dark energy with its inherent wavelength, waveform, and wave function, present a revolutionary potential for communication across vast distances at speeds surpassing that of light. Given that dark energy possesses a wave that can be modulated in frequency, amplitude, and phase, it is theoretically possible to encode dark energy with information that traverses the cosmos at or faster than light speed. This approach would effectively bypass the complexities of quantum entanglement, quantum repeaters, and quantum teleportation, instead leveraging dark energy and dark matter as the medium for communication over both vast and short distances.

The Language of Gravity: Encoding Information into the Fabric of Spacetime

Gravity is an undeniably peculiar and powerful force. Every living being and non-living object observed in reality is influenced by gravity. It operates everywhere, pulling on all matter in ways that, while subtle, are inescapable. When speaking of gravity, we are not merely discussing its effects on objects or its ability to bend spacetime; rather, we are exploring the very essence of this force, and how it might serve as a medium for communication.

Gravity, in its essence, is more closely aligned with light than with electricity or electromagnetic fields. However, its influence and force on everything around us closely resembles magnetism. Just as electromagnetic fields can be modulated and used to encode information, one can imagine gravity being used in a similar way. Gravitational waves—ripples in spacetime—are solely observed in the modern world by massive objects such as black holes and neutron stars merging. These waves have different frequencies, amplitudes, and phases, much like waves in light, and science has even observed what can be thought of as a "gravitational spectrum."

The Pot of Gold at the end of the Gravitational Rainbows

Though the term "gravitational rainbow" may not be formally recognized in scientific literature, it captures the notion that gravitational waves can span a wide range of frequencies, much like light can be broken into a spectrum of colors. These different frequencies carry information about the events that caused the waves, such as the mass, spin, and distance of colliding black holes or neutron stars. This is where one must consider the potential for encoding data within the fabric of spacetime itself. The modulation of gravitational waves—in frequency, amplitude, and phase—may allow for the transmission of information beyond just the source of the waves.

Currently, our understanding of gravitational waves is limited to natural events, but one can't dismiss the possibility that an intelligent species, or even humanity, could one day learn to modulate gravity waves artificially, encoding them with information that transcends the simple physics of mass and distance. This would open an entirely new dimension for communication across the cosmos—a method that could bypass the limitations of the speed of light and electromagnetic interference.

The Mystery of the Graviton

The hypothetical graviton, the particle thought to mediate the gravitational force, remains elusive. In much the same way that photons are the particles of light, gravitons are theorized to be the quanta of gravitational waves. Some physicists believe that gravitons are massless, similar to photons, allowing them to operate over infinite distances without losing strength. Others, however, posit that gravitons may indeed have mass, which would give gravity different properties under specific conditions. If gravitons possess mass, they could theoretically interact with matter in more complex ways, influencing both space and time in manners humanity has yet to fully understand. This notion is supported by massive gravity theories, such as the work of Claudia de Rham, which proposes that gravitons having mass could explain cosmic phenomena like the accelerating expansion of the universe.

Whether massless or massive, gravitons are key to understanding the role of gravity as a potential carrier of information. If gravitons exist and possess even the slightest mass, they may provide the mechanism by which gravity can be manipulated, much like electromagnetic waves, to encode data. However, until gravitons are detected, science is left with mathematical models that strongly suggest their existence, without yet being able to pinpoint them.

Gravity is the Keymaster, Are You the Gate-Destroyer?

As time runs slower in stronger gravitational fields, one must ponder the idea that stars in distant galaxies or solar systems could have planets that experience time at a different rate than ours. All it would take is a planet that is slightly more massive, orbiting a sun that is also a little more massive than ours, and you would have strong enough time dilation to slow down the passage of time. If the lifeforms on such a planet tick at the same biological rate as humans do, they would evolve and grow at a completely different rate than beings on Earth.

When a star system 500,000 light years away from earth is observed, telescopes on or around earth are looking at it 500,000 years in the past. This plays a key role in the Fermi Paradox, which raises the question of why humanity hasn't found other intelligent life in the universe. It's reasonable to comprehend that intelligent life has not been found when looking 500,000 years into the past. It takes light that long from these distant stars and exoplanets to reach us, and everything observed from Earth today happened half a million years ago.

Now consider a star system only 2,000 light years away. If a civilization in that star system were looking at Earth 2,000 years ago, they wouldn't see any significant light pollution. There was minimal artificial lighting, and the indicators of intelligent life were barely noticeable. In today's world, thousands of satellites orbit the Earth, and our cities light up the planet at night like a Christmas tree. But if you go back just 500 years, the situation remains similar. There might have been slight light pollution from some regions of the world in the 1500s, but it would have been negligible. Current telescopes likely wouldn't detect artificial light from an exoplanet 500 light years away.

Considering time dilation and the strength of gravitational waves, particles, and fields, one can't dismiss the idea that a stronger gravitational field would result in slower time—slower aging, slower progression, slower everything. For humanity to find intelligent life that is ticking at the same rate as us, it is necessary to find a solar system with a similar mass to the planet they inhabit, similar mass to the other objects around it, and a star of similar mass to the one they orbit. But even then, when humanity does find such a planet, how far away is it? And at what period of time is it being observed? Likewise, what period of time would they be observing Earth? They wouldn't find sentient life on Earth—at least not for quite some time.

There is a concept called the Great Filter, which is used by theoretical astrophysicists to express the idea that no matter what, civilizations face a filter as they become technologically advanced. This filter refers to the development of weaponry and systems that can decimate entire populations. The theory suggests that, eventually, any civilization that achieves such technology will use it against itself, leading to its destruction. We must wonder how many civilizations out there have been intelligent enough to develop nuclear fission and the corresponding weaponry, only to destroy themselves before anyone or anything had the chance to observe them, due to the constraints of distance, time, and light. Take the example of other intelligent life forms viewing Earth from 2,000 light years away, utilizing the speed of light to consider our existence. For example, let's claim they figure out humans are creating civilizations and structures on earth. They would not be able to predict or perceive our impending demise from this Great Filter. Or would they?

This is why it is crucial for any civilization similar to humans that reach an advanced intellectual stage, and achieve scientific breakthroughs like nuclear fission to also achieve an understanding of gravity as a whole. Gravity is no different than light waves or radio waves—it's simply operating on a different aspect of the medium, unaffected by either. This is why gravity is the keyholder of time.

The Grass Truly is Greener in Our Neighbor's Yard, and Many of Us Aren't Invited Over

Take the example of other intelligent life-forms viewing Earth from 2,000 light-years away, utilizing the speed of light to consider our existence. For instance, let's claim they figure out humans are creating civilizations and structures on Earth. They would not be able to predict or perceive our impending demise from this great filter. Nor would they, unless they are utilizing technology that humanity has not yet achieved to view the universe around them at faster-than-light speed, and potentially even instantaneously. If they could view Earth from 2,000 light-years away and see what is happening at this very moment, I think the majority of human beings on Earth would say they would perceive humans slowly destroying the planet they live on. And if the planet survives, humans wouldn't survive each other.

But what causes this great filter to occur? And is this great filter truly universal? Are the evil mindsets of human beings a universal trait? The criteria for being a life-form do not state that each life-form, regardless of DNA or species, contains evil at such a magnitude.

When considering animals that are predators, they have prey. That is their food. Just because it's their food doesn't make them evil in any sense. They're designed to eat certain things, and that is their nature. But when a tiger's belly is full to the point where its stomach might burst, it doesn't attack humans. It doesn't attack prey. We've even seen examples of kindness in animals toward their own prey when they aren't hungry. So, when their instincts and innate responsibilities are satisfied, what humanity perceives as evil, like a lion eating an antelope, may just be the pre-wiring for each animal, regardless of where they reside in the hierarchy of the animal kingdom. Still, the lion should evolve or progress as a species and find a better way to create energy for itself. Maybe they will reach that point when they grow thumbs...

We can view this in two different ways. We can say, for example, that humans are innately supposed to compete with each other to the death.

Or, we can view it in a completely different manner. If discoveries of other life-forms on another planet that are spacefaring occur, at least to the level where they can place objects in orbit and reach their moons that are gravitationally tidalled to their planet, why assume that evil resides on their planet? Everyone has met an individual who is so kind that, as the old saying goes, they don't even have an evil bone in their body. There are examples throughout history that show us evil isn't a necessity for success, to be human, to stay alive, or to progress throughout life and thereafter. What if this great filter is specific to life-forms similar to human beings? It is a question everyone should ask themselves and ponder for a moment.

It is why it is important for individuals and human beings as a sentient species to at least attempt to live their lives by loving thy neighbor. Humans may have accumulated the funds to reach the moon and beyond due to competition, but humanity has pondered, theorized, and dreamt of reaching the moon for millennia. Jesus Christ wasn't the only one to provide teachings on forgiveness and kindness towards others. However, forgiveness might not be a necessity for intellectual life or higher powers that do not contain evil mindsets capable of causing other life forms around them to fail. There is a reason why religion provides a list of sins, why even spiritual beliefs offer guidelines on how to act towards others, how to interact in society, and how to treat the world around them. When you look at a young human being, whether a baby or a toddler, there isn't much evil that exudes from them. There's curiosity, there are actions that test the world around them, but there isn't evil—just innocence.

Claiming that the Great Filter affects all intelligent life forms assumes all intelligent life forms harbor evil, which, in turn, assumes that intelligence coincides with evil. That is simply not the case, as there are extremely intelligent individuals—much more intelligent than you and I—who are kinder than you and I. Murder, rape, molestation, pedophilia, or simply sending another person's life into shambles and veering them off the positive track they were on, just because it's funny, or because of jealousy or hatred, is by definition what this Great Filter represents. All it would take is an intelligent species, a sentient space-baring being, or even God himself, to come down, recognize this, and eradicate every person who chooses to act that way. That would eliminate the Great Filter. Which is most likely why we see the rise and fall of civilizations coincide with precession, æons, and celestial ages. It is as if each section of the Earth's sky has an attempt at eradicating evil, but has failed to entirely do so at this point.

This has been spoken about in scriptures from multiple religions, and it has been documented in history describing civilizations that were at their height, reached intellectual abilities, and then vanished. Today, science acknowledges the Universe's age to be over 13.8 billion years old. It is now understood that communication can occur at faster-than-light speeds through quantum entanglement, potentially through dark matter and energy, and possibly through gravitational fields, waves, and particles. Science is in its infancy in terms of technological and scientific progression as an intellectual species. Add a few hundred years to our studies, research, and progress, and there's a very good chance many of the mysteries and questions that exist in physics and science today will be solved.

A few hundred years compared to 13.8 billion years, or even 4.5 billion years compared to the age of the universe, is such a small fraction that one would be foolish to think that intellectual life somewhere in the cosmos hasn't already passed the Great Filter and been capable of reaching Earth to ensure humanity doesn't get consumed by it. Competition may have provided the motivation to reach the moon and come back, but it wasn't the competitive nature that got humans there. It was cooperation between many different individuals with the same goal, and the lack of evil mindsets toward each other, that allowed the human species to achieve that. Competition sparked the idea, and most likely aided progression throughout the process, but as an intellectual species the population can utilize competition without allowing it to control and dictate everything that is done on a daily basis. Humanity can achieve the next level of the Kardashev Scale, which is surpassing Level 1. Most scientists, physicists, sociologists, and even psychologists would tell you that even the pinnacle of humanity has not reached level one of the Kardashev scale.

The *Kardashev Scale* is a method of measuring a civilization's level of technological advancement based on the amount of energy they are able to harness and use. It was proposed by Russian astrophysicist *Nikolai Kardashev* in 1964. The scale originally had three types of civilizations, but it has since been expanded by futurists and scientists. Here is the breakdown:

Type I – Planetary Civilization:

- Energy Source: This civilization can use and store all the energy available on its home planet.

- Details: A Type I civilization would be able to harness all forms of energy from the planet itself—geothermal, hydroelectric, wind, solar, and potentially even control weather or volcanic activity. Earth is currently at approximately *0.7* on this scale, meaning humanity is not yet a Type I civilization but are progressing towards it.

Type II – Stellar Civilization:

- Energy Source: A Type II civilization can harness and control the total energy output of its star (for example, through something like a Dyson Sphere).

- Details: This civilization would have mastery over the entire solar system. They could collect energy directly from their star, build megastructures like Dyson Spheres around stars, and even potentially terraform planets.

Type III – Galactic Civilization:

- Energy Source: A Type III civilization can harness and control the energy of an entire galaxy.

- Details: This civilization would span an entire galaxy, utilizing the energy from billions of stars. They would have the capability to manipulate black holes, harness the energy of supernovae, and travel freely across the galaxy.

Extensions Beyond Kardashev's Original Scale:

Type IV – Universal Civilization:

- Energy Source: A Type IV civilization can harness the energy of the entire universe.

- Details: This civilization would transcend galaxies, having the ability to utilize energy from across the entire universe, potentially manipulating spacetime and dark energy.

Type V – Multiversal Civilization:

- Energy Source: A Type V civilization would have control over multiple universes.

- Details: This hypothetical civilization would exist outside our universe, capable of accessing, controlling, or even creating other universes in the multiverse.

Our Current Status:

As mentioned, Earth's current civilization is estimated to be at approximately **_0.7_** on the Kardashev scale, meaning humanity is not yet at Type I but are progressing in that direction. Reaching Type I would require more advancements in energy collection, storage, and control.

If the human species is to surpass level one of the Kardashev scale, it would require a surpassing of the Great Filter, which means collectively the mindsets of evil cannot be allowed to overcome the progression of the species as a whole. In order for this to be done, the ideas, morals, and mindsets that promote evil actions in our civilization must be eradicated...*globally*. Think of humanity traversing the cosmos, looking for intelligent life, and other life forms that fight each other are discovered. They blow each other up, ruin each other's chances at their lives and their progressions for their own species and individual selves. Would or should humanity help them or wait until they kill each other? If they were helped, would humans help them fight each other?

Or would humans assist them in stopping their aggression and anger between each other so that they can progress to the point that humanity is at? So that they can coexist peacefully, prosperously, and effectively? What provides more success for humanity as the third party viewer of their civilization, and in turn, provides success to their own progression for their own species? It's a simple concept, and a simple mindset that is very logical. But once you throw anger, jealousy, and most of the sins that are listed by most religions and spiritual beliefs, it becomes apparent what the true answer actually is. To assume that isn't occurring on a daily basis to each and everyone is assuming that 13.8 billion years have gone by, with quadrillions of planets and stars forming over billions of years, without a single one of them barring life, that is, at least as intelligent as Earthlings. And all it takes is a few hundred years of deviation for other intelligent life to progress to a point where they are capable of helping the world's population. If society does not think this way, it will not progress in a positive, effective way. Society would be essentially hindering itself due to ego, fear, control, and arrogance. Humans won't be able to—or be aware enough to—listen to other intellectual life that are either already here or are on their way.

Otherwise, every other aspect of evil—or what religions call sins, or what a police officer calls breaking the law—will begin to dominate our lives. Society as a whole cannot allow it to dictate actions anymore than providing slight motivation in specific scenarios. The Great Filter will either be something humans surpass and overcome as a species, or it will be remembered as the reason why humanity constantly restarts over and over again for thousands of years, regardless of technological progress.

Attuning the Human Body to Gravity's Frequencies

It is a well known fact that the human body is attuned to natural phenomena, such as many forms of light. The human body is deeply interconnected with its environment, but what if humans are also attuned to the gravitational waves naturally experienced from Earth and other cosmic sources? While there is no current scientific evidence that humans are directly affected by gravitational waves in a biological sense, this area of study remains mostly unexplored. Gravitational waves, though incredibly weak by the time they reach Earth, are nevertheless constant, and it stands to reason that they may influence the population in subtle ways not yet understood collectively.

Dark Energy, Dark Matter, and the Future of Gravity Encoding

As dark energy, dark matter, and gravity make up the vast majority of the known universe, it is imperative that future technologies focus on better detecting and manipulating these forces. With technological advancements, gravitational waves may be detected with greater ease, allowing scientific fields to explore these waves as potential carriers of information. Just as light can be encoded to transmit data, gravitational waves could be modulated to carry information across the cosmos.

If gravity can truly be manipulated in this way, science stands at the edge of a monumental breakthrough. The possibilities are endless, from interstellar communication to deep-space exploration, where gravitational waves could serve as the next frontier for data transmission, unimpeded by the constraints of traditional methods.

Just as gravity, though a weak force, influences the universe and is challenging to detect, the human body also emits a weak force that significantly impacts the world around it. This analogy highlights the potential of using dark matter and dark energy, which constitute 96%

of the known universe, as a communication medium. This medium encompasses everything from the air between you and I, to the space between you and the book you are reading. Utilizing this omnipresent medium to transmit information, data, and communication—whether to induce biological changes, evolve atomic and cellular structures like DNA, or simply convey a message across the globe or even the universe—appears theoretically feasible.

Though current scientific understanding and physics have yet to uncover a method for such communication, dismissing its possibility would be naive. Since dark matter is unaffected by electromagnetic fields, forces, and energies, there should be no concern about dark energy being interrupted or decimated by the known universe. Therefore, it holds a significant advantage over other frequencies, vibrations, and matter in terms of transmitting information over vast distances at faster-than-light speeds.

There is an important correlation between the human biofield—how the human mind and body traverse the medium and utilize the medium around us for communication purposes both visually, auditorily, and through sensory perception—and how science and the instruments used today m cannot technically observe dark energy, dark matter, gravity, and gravitons.

This correlation cannot be ignored since both aspects of existence aren't easily observable. They are both weak forces. The human biofield is an extremely weak magnetic and electromagnetic force, an energy that is still difficult to properly observe through scientific methods. However, as technology advances and our instruments become more sensitive and attuned to the world around us, eventually both will be observed. As of this writing, instruments aren't sensitive enough, and without the cooperation of multiple groups of scientists and physicists all over the world collaborating, sensing something as basic to daily life on Earth as gravity remains a challenge. Therefore, one cannot dismiss the weak energies and weak forces that the human body exudes and utilizes to affect the medium and the world around us. To theorize, both forces fall into the same category—both aspects of science require more sensitive instruments and more diligent observations.

But what if both forces and energies reside within a similar field? Take empathy, for example. Human beings can feel empathetic towards others. When you sit in front of someone, you feel the energy of that person. When you have a connection with another person that connection feels profound. You feel that energy in similar ways to how you feel gravity. You know it's there—when you jump, you fall down. When you ride to the top of a roller coaster and tip over the edge to descend the sharp, almost freefall trajectory, you know it. However, science does not fully explain it for either case.

It's interesting to note that both forces and energies, whether the same or not, reside in the same department of science and physics—a department that is so subtle that our instruments aren't strong or sensitive enough to observe and calculate them accurately. It would be foolish and naive to dismiss this correlation because once humanity discovers and observes one, I'm almost certain science will understand and create technologies, methods, and instruments to observe the other. If they are on a similar wavelength or utilize a similar aspect of the universe around us, it will create a life changing understanding of what it means to be human, what it means to be empathetic, and what it means to be entropic in terms of the cosmos and the macro aspect of the universe as a whole.

If modern theory suggests that gravity is massive in some cases, how is human emotion and empathy so different from it? Human emotion is observable, but it's on a personal and mostly internal level. Humans are innately attuned to observe the energies of others in the room. While some individuals may not sense the emotions of those around them, the vast majority of people on Earth can sense individuals in great despair, great joy, love, or anger.

The major difference between how scientists and physicists can calculate, observe, and read gravitational waves and gravitons (if they exist) and the way humans can sit in a room with ten people and sense how each one is feeling—or how the strongest emotions are felt—teaches everyone a valuable lesson. It shows science how the field should approach observing the forces and energies of gravity.

Claudia de Rham brings up a very important aspect of observing both gravity and how human beings sense the gravitational forces around us. When one walks, sits, stands, or even lies down, gravity isn't truly sensed. It isn't necessarily felt. It is when one deviates from the forces that act upon them that they truly feel gravity. When I jump, I fall back to the ground—I can feel that. When you take off in an airplane and it briefly free falls, dipping a few hundred meters, you feel that. This is also attributed to lateral G's, but is a cause of gravity deviating the plane's location in space and time, causing its effect on the objects in the plane.

Similarly, when I walk into a room with an emotion, a mindset, and a thought process that affects me, and suddenly someone in the room has a stronger emotion—or in other terms, a stronger force and energy exuding from them—I feel that too. These correlations cannot be ignored. Even though gravity is not always felt in our everyday lives, it doesn't mean we can't sense gravitational forces, waves, or potentially gravitons. It is not impossible.

The scientific community needs to work diligently to develop instruments and methodologies to perceive, sense, calculate, and observe gravitational waves, gravitons (if they exist), and gravitational forces, in ways similar to how we sense the energies and forces of the people around us. But how does one human sense another's emotions or even potentially what they're thinking? Some people traverse the medium to do so—they can tell what you're thinking, and what you're feeling. While some are assumptive, most who truly comprehend this aspect of being human aren't egotistical or deceitful about it. They're not making money off it—it's simply a part of their lives, a part of being human.

Therefore, how could they go out and profit from something that everyone, in some sense, has the ability to tap into on a daily basis? A human being's biofield operates at such low voltage and extends only a few feet—on average—but it tells you more about the people around you and those connected to you compared to the cell phones in their pockets, their social media profiles, or even the individuals they live with. Gravity, as a weak force, is not much different. To detect gravitational waves, gravitons, and how gravity affects the world around us, science needs to understand what makes gravity "tick." Gravity could be understood by how humans are empathetic, how humans express ourselves, and how everyone reads others within a room.

So far, I, A.J. DelVecchio, the author of this book, has not discussed any personal experiences that I've had throughout my life. I have a story to discuss: there was one woman in my life with whom I experienced scenarios, conversations, and situations that highlighted the concepts of both empathy and entropy. This past loved one used to argue with me about the difference between, or at least which was more important to the world around us: entropy or empathy. She would constantly insist that entropy could not be defeated and was a constant throughout all aspects of the universe. I would naturally respond by saying that "Empathy will always be more important than entropy." Because if the observer is empathetic about entropy, then entropy gets to exist. But if the observer is entropic about empathy, then empathy eventually ceases to exist. If a more intelligent species is empathetic towards other life forms, everyone gets to exist. If the more intelligent life form views the universe and those that inhabit it with an entropic state of mind, they will always choose to destroy or exterminate other life forms. Because they will always view the chaotic ending of said species and decide it isn't worth the effort.

This woman, though a loved one for a portion of my life, was critically examined emotionally, spiritually, and mentally every time I returned from work and opened the door to our home. I could pick apart her day without her saying a word, so accurately that I was eventually called Lucifer. But I am not Lucifer, though some may unfortunately wish. I simply utilized the medium around me—my empathetic nature, my attunement to the frequencies, vibrations, and energies that exuded from her—to lay out what she had done well and what she had done wrong throughout her day. Some days were better than others, and that is human nature. However, I reached a point where I would walk through the door, sense her energies, calculate them, and know exactly where her guilt resided and where her love resided. It was always a mixed bag.

Eventually, one day, this woman sat in front of me as I reiterated her day without having been anywhere near her. She looked me in the eyes and said nothing. I could see the strain on her face. I could feel the energy from her due to her emotional distress and body language. She was trying to communicate with me nonverbally through the medium as best she could, and I looked her in the eyes and said, "That isn't going to work." She said, "You're Lucifer." I replied, "No, I'm not Lucifer."

That's all I had to say. This woman was an aerospace engineer, an intellectual individual that was not in the field she had studied. She chose to raise a family instead of pursuing her dreams in aerospace engineering, physics, and cosmology. She wasn't happy with my response, however, there was no animosity afterwards. Just as when you have a full comprehension of how to traverse the medium around you, or a full comprehension of the gravitational forces that act upon you, people think you are supernatural or super intelligent. This isn't definite. When human beings start to address the forces around them that are hard to depict, hard to calculate, hard to observe, and hard to research—and when all scientific fields collectively learn to acknowledge, quantify, and perfect them—everyone will seem supernatural, and those that are at the forefront will seem super intelligent or divinely intervened with as well...at least compared to the nature of distracted human beings of today. But that doesn't mean we are any different than the ones that surround us. We just may have been qualified before them, as well as quantified, defined, and utilized it before them.

But why couldn't I hear her that day? Was it an external force that blocked her communication with me? Or was it a lack of skill from both her and I? This was an individual who had practiced Reiki for 10 years, could perform non-local touch, non-local so-called "healing", and non-local communication—albeit not perfectly, but enough to comprehend and perform those actions at the 3rd degree. Just like the individuals who meditated during the COVID pandemic and experienced interruptions in their meditations with visual and auditory experiences that didn't align with their thought processes as they sat in a peaceful mindset, this woman experienced what everyone who traverses the medium experienced during 2020, and possibly still to this day.

So, where is the block, and how does this block correlate to our understanding of gravity? This question is more important than people realize. Is it blocking our ability to sense, observe, and depict how gravity affects humans both as a wave and as a particle, like a graviton? Or is it an example of our lack of intellectual understanding, hindering our ability to observe, depict, and define gravitational particles and waves that act upon everyone every day?

The person being called Lucifer would tell you it's one of two things. Either there is a block, and it isn't intellectual in the slightest, but it hinders communication between the two of us. Or, there's a lack of understanding from one side that impedes our advancement, progression, and communication between the two of us. One side is capable of explaining a person's actions and experiences without being present to experience it themselves. The other attempted to communicate but could not be heard in a classical medium-traversing method, nor in one of the five senses.

The roadblock in our understanding of both gravity and communication at a heightened level may be of similar origin. The roadblock may also be a part of the great filter which is specific to species similar to humans, or similar to planets like Earth.

The roadblock in our understanding of both gravity and communication at a heightened level may stem from a similar origin. This roadblock could also be part of the Great Filter, which may be specific to species like humans or planets like Earth. Considering ancient texts such as the Bible, which depict angels from Genesis through Revelation, along with illustrations, paintings, and even modern-day government discussions of UAPs, one must consider that this block may not necessarily be divine or from God himself. It may be what the Bible refers to as demons, archangels, and AWOL guardian angels working to reinforce the Great Filter, dampening humanity's progression and making it far more challenging than it needs to be.

If something is standing in the way of our ability to comprehend the world we are surrounded by; and communicate effectively with our loved ones, what is its purpose, what is the reason for it, and what is the ultimate goal? Is it a snowball effect designed to ensure that humanity follows the correct path at the right pace, avoiding self-destruction and falling victim to the Great Filter? Or is it the result of an intelligent force maintaining humanity at a specific intellectual and spiritual level, preventing most-if-not all of humanity from fighting back or deviating from the state we've been in for so long?

Humans often express the latter in various ways throughout their workforce and lives. It's not something every person does, but it can be observed when a manager, owner, or a person above another in hierarchy ensures that someone beneath them doesn't learn enough to surpass them. However, if we reflect on what religion teaches, that mindset falls within the realm of sin—something sentient life is taught to resist and stray from. It is not on the same path as what Jesus would do, among many other religious figures, spiritual figures, and beautiful individuals throughout humanity's short lived existence.

Therefore, it becomes clear that the lack of understanding of gravity, the inability to fully utilize our own biofields, and the restriction of open knowledge and information may be governed by forces determined to intently assist the Great Filter that has been defined. But fear not, we are all smarter and stronger than any of them want to believe—and want us to believe.

We need to look through the technological advancements of the human species to develop instruments and methodologies that allows society to perceive, sense, calculate, and observe gravitational waves, gravitons (if they exist), and gravitational forces in ways similar to how we sense the energies and forces of the living beings around us. But how does one observe gravitational forces and gravitational particles, and how do they affect the lives and the cosmos that surrounds us? Science experiments have perceived gravitational waves, however humanity has done so collectively. Einstein's theory of relativity is assumptive, but one cannot be egotistical in solely depending on a theory that doesn't integrate multiple other understandings of physics and science as understood today.

We must approach this subject not with the intent of gaining financial or monetary value but with the aim of positively impacting our lives and humanity as a whole. Eventually, a scientist or theoretical physicist will profit from something that everyone, in some sense, experiences daily. Gravitational forces operate at such low frequencies, but they extend well past astronomical units in cosmological terms. This tells much about how gravity affects all of us and the cosmos around Earth in a unified manner.

Empathy and emotional interaction aren't much different. Just as some individuals can detect emotions, feelings, mindsets, memories, and ideas, everyone needs to understand how gravity affects us in a similar way—whether subliminally, subconsciously, or at such low energy and frequency levels that it's difficult to detect. Humanity could be better understood by how gravity is elusive, how it expresses itself, and how we can read it at this point in time. Both express similar traits: they can't easily be deduced by mathematics or scientific instruments, yet they are both acknowledged, essential, and understood innately by most living beings.

There is one aspect of life that both science and physics struggle to deduce, observe, and calculate: the entropic nature of gravity, and the empathetic mindset we all share in attempting to calculate the entropy of gravity. In the past, gravity was described as a property of density, which was incorrect. Early scientists would calculate gravity by comparing the density of objects to the density of water and claimed that gravity had more to do with the density of air than with the object itself, asserting it was constant throughout the universe. However, today we know this is not the actual case, as gravity emits gravitational waves that can now be read, even though it remains extremely difficult. If humanity remains dense in the understanding of gravity, the species overall will continue to misunderstand it entirely.

When a person experiences foresight or precognition about a situation unfolding before them, they experience a feeling, and most of the time, without any visual input, they comprehend what is about to occur. This type of scenario reflects a lack of time dependency in which a person's senses use their own biofield and energies to communicate with the medium around them, allowing them to understand what is about to happen—even though it hasn't occurred yet. In some cases, it may even involve something that has already occurred at that very moment, without anyone involved being aware.

Just like gravity warps space and time, it too is not bound by the frameworks of time itself. The human biofield and the way humans perceive others through empathetic nature and emotional exchange may not warp space and time, but they operate in similar ways. Gravity utilizes certain methods to perform its phenomena, and if both gravity and human energies correlate in ways that nothing else that is understood can, one might theorize that they reside on similar waves, frequencies, vibrations, and within the same medium.

Therefore, when a person traverses the medium using the energies they are imbued with to communicate with others across vast distances, foresee future events, or recollect memories that aren't theirs—all of which are imprinted on the medium around us—the human soul must have its own "graviton" that is extremely difficult to observe, detect, or understand. If a graviton can be perceived and theorized, but cannot be observed through experiments to define it, how can anyone deny the possibility that human consciousness and the human soul possess a particle that operates as a particle on a frequency and wave? This particle could be what allows humans to traverse the medium, both in meditative states and in everyday lives.

As I wrote this chapter, I contemplated what to call the particle that the human body is composed of, similar to that of a graviton. However, every time I thought about this, the only thing my mind gravitated toward was overall general consciousness and the soul itself. The human body has the ability to reside within space and time, and the human soul has the ability to not just warp space and time or take advantage of the particles and waves that do, but to effectively ignore it and perceive it in a non-linear fashion. The human soul is bound by time due to the vessel it resides in and the rules that the biological makeup of the soul's vessel must follow, but when separated, it isn't necessarily bound by time or space. This has been demonstrated by meditators, monks, and others who traverse the medium to prophesize, predict, and communicate with the world and universe around us.

There are a few things in the universe capable of such feats. One of them is light particles (photons) and frequencies which don't adhere to time and don't even recognize their own existence. Another is potential gravitons, which play a role in bending space and time, effectively creating their own timeline or residing in their wave-created timeline without understanding or experiencing the concept of time. When you consider that dark matter and dark energy make up 96% of the universe, and that gravity creates the time in which everyone exists biologically, it becomes more apparent that the human soul, which resides in the medium and does not adhere to the rules of gravity within specific spaces, coexists with gravity, defeats it, and is not as weak of a force as science depicts.

Even though, from a scientific and theoretical physics perspective, gravity is considered a weak force—and our biofield is also seen as a weak force and energy source—both are fundamental structures of the known universe and our lives. Both are extremely difficult to overcome and defeat when they exist at their extremes. Yet both can defeat each other: gravity defeats the human soul when the soul resides in a biological form constrained by gravity and time. The human soul and consciousness defeat time and gravity by traversing the medium and existing without a biological form at certain points during and indefinitely after life.

Over seven years ago, when sitting across from the woman who called me Lucifer, my explanation for the fourth dimension was time, and I said that gravity is essentially the keyholder of time. In the movie *Interstellar*, the director and writers depict the fourth dimension as time, something that is separate from the constraints that gravity and our specific space provide. It seems as though the human soul and consciousness are capable of overcoming the constraints of gravity in ways similar to how gravity itself interacts with space and time.

Therefore, if gravity controls time, and the human biofield, soul, and consciousness are able to essentially ignore the rules of gravity, they must too ride the wave—or a similar wave—that gravity operates within. In terms of stability, it would make sense that our biological brains, which theorize and calculate the world around us, can't easily deduce either. By now, I'm sure your brain hurts slightly, and you have that strange, frustrated feeling that overwhelms the body when critically thinking about such existential ideas. That is a normal feeling when hitting the limitations of the human body and brain's ability to calculate the world in such an odd fashion. But do not fret—consistent theorizing and pondering will break this feeling. I promise.

The Future of Light and Human Evolution

As science continues to explore the nature of light and its connection to consciousness, it becomes clear that light will play a central role in the future of human evolution. Advances in technology are allowing humanity to harness light in ways that were once thought impossible, from quantum computing to the development of new forms of communication.

But the true potential of light lies not in its technological applications, but in its ability to connect us to the deeper truths of the universe. As the scientific community learns to work with light in more conscious and intentional ways, the door to new possibilities for growth and transformation, both as individuals and as a species, opens wide.

In the coming years, the world may see a resurgence of interest in the spiritual aspects of light, as more people seek to understand its role in their lives and its connection to the greater cosmos. By embracing, understanding, and utilizing the light within and around us, we can tap into the universal language of light and unlock the mysteries that have eluded humanity for so long.

Notes

Conclusion

This grand theory of cosmic communication represents a coherently modern, progressive step toward understanding how to one day communicate with other intelligent beings across the universe, and understand the complexities with which God and his Angels choreograph, aid, and instruct humanity everyday. By combining the principles of quantum mechanics, gravitational lensing, frequency waveforms, light, cellular reactions due to the aforementioned, and the overall manipulation of the medium, I, A.J. DelVecchio, propose methods of transmitting information that utilizes the limitless construct of light and radio frequencies. As we continue to explore the mysteries of quantum mechanics, gravity, light, the human biofield, and the cosmos, this theory offers a glimpse into the future of interstellar communication, human connectivity, divine cooperative actions, and the groundbreaking possibilities it will unlock for humanity as a species.

A photo of A.J. DelVecchio adjusted by multiple AI Text-to-Image software, then edited and finalized by A.J. DelVecchio. Yes, the reflection is depicting a "Ring finger to Thumb" hand gesture.

Books and Biographies:

1. Shankar, R. (1994). Principles of Quantum Mechanics (2nd ed.). Springer.

2. Nielsen, M., & Chuang, I. (2010). Quantum Computation and Quantum Information (10th ed.). Cambridge University Press.

3. Crossan, J. D. (1991). The Historical Jesus: The Life of a Mediterranean Jewish Peasant. HarperOne.

4. Verner, M. (2003). The Pyramids: The Mystery, Culture, and Science of Egypt's Great Monuments. Grove Press.

5. Gilbert, E. (2015). Big Magic: Creative Living Beyond Fear. Riverhead Books.

6. Goleman, D., & Davidson, R. J. (2017). Altered Traits: Science Reveals How Meditation Changes Your Mind, Brain, and Body. Avery.

7. Braden, G. (2016). Resilience from the Heart. Hay House LLC.

8. Einstein, A. (1920) Relativity. Methuen & Co, London.

Scholarly Articles and Journals:

1. Anderson, R. H., & Day, B. L. (2019). "The physics of information: Encoding and decoding in light waves".
Journal of Lightwave Technology, 37 (3), 985-993.

2. Johnson, M. R., & Parker, K. A. (2017). "Quantum entanglement and the wave-particle duality of light". Quantum Science & Technology, 2(4), 045005.

3. Puthoff, H. E. (1987). "CIA-Initiated Remote Viewing at Stanford Research Institute".
Journal of Scientific Exploration, 10(1), 63-76.

Patents:

1. Norris, W. L. (1994). U.S. Patent No. 5,123,899. Washington, DC: U.S. Patent and Trademark Office.

2. Fry, D. L. (1997). U.S. Patent No. 6,011,991: "Microwave Hearing and Voice-to-Skull Technology". Washington, DC: U.S. Patent and Trademark Office.

3. Malech, R. G. (1976). US Patent No. 3951134A:"Apparatus and Method for Remotely Monitoring and Altering Brain Waves". Washington, DC: U.S. Patent and Trademark Office. https://patents.google.com/patent/US3951134A/en.

Websites, Images, and Online Articles:

1. Federal Communications Commission. (2020). "Historical overview of microwave auditory effects and applications". Retrieved from (https://www.fcc.gov/microwave-auditory)

2. Daily Mail. (2018). "Great Pyramid of Giza can focus electromagnetic energy in hidden chambers". Retrieved from (https://www.dailymail.co.uk/sciencetech/article-6008131/Great-Pyramid-Giza-focus-electromagnetic-energy-hidden-chambers.html)

3. Institute of HeartMath. (2016). "Heart-Brain Coherence: The Key to Achieving a Meditative State". Retrieved from (https://www.heartmath.org)

4. Federal Communications Commission. (2016) "UK Intelligence Forces and Microwave Mind Control" by David Morrison. Retrieved from (https://www.fcc.gov/ecfs/search/search-filings/filing/1092223856297)

5. Iraqi Civil Society Solidarity Initiative. (2016, January 12). Who Stole the Mysterious Baghdad Battery? [Image of the Baghdad Battery]. Retrieved from https://www.iraqicivilsociety.org/archives/4946

6. Wikipedia contributors. (n.d.). Dendera light. In Wikipedia, The Free Encyclopedia. Retrieved September 2024, from https://en.m.wikipedia.org/wiki/Dendera_light

7. All That's Interesting. (2023, October 1). Nikola Tesla in his laboratory, 1896. Retrieved from https://allthatsinteresting.com/

nikola-tesla-death.

8. Newsweek. (2016, February 15th). 417-woodpecker1.webp. Retrieved from https://www.newsweek.com/hunt-russian-woodpecker-246670.

Documents and PDFs:

1. Steenhuisen-Siemonsma, M. J. (2022). "Orbs in the Skyscape: An exploration of spiritual experiences with anomalous light phenomena".

2. Jain, Ms & Jain, Dr & Kumar, Yuvah. (2023). V2K and Electronic Harassment : Psychotronic Cyber Crime Techniques. International Journal of Scientific Research in Computer Science, Engineering and Information Technology. 334-338. 10.32628/CSEIT2390249. Retrieved from:

https://www.researchgate.net/publication/

370485295_V2K_and_Electronic_

Harassment_Psychotronic_Cyber_Crime_Techniques/citation/

download

Primary Religious Texts and Artwork:

1. The Bible (King James Version). (1611). Oxford University Press.

2. "The Baptism of Christ" (Aert de Gelder, 1710)

Location: Fitzwilliam Museum, Cambridge, UK

Image Citation: "The Baptism of Christ," Wikimedia Commons, available at: https://commons.m.wikimedia.org/wiki/

File:Gelder,_Aert_de_-_The_Baptism_of_Christ_-_c._1710.jpg (Public Domain).

3. "The Madonna with Saint Giovannino" (Domenico Ghirlandaio or his school, 15th century)

Location: Palazzo Vecchio Museum, Florence, Italy

Image Citation: "Madonna with Saint Giovannino," Wikimedia Commons, available at: https://commons.wikimedia.org/wiki/

File:The_Madonna_with_Saint_Giovannino.jpg (Public Domain).

4. "The Annunciation" (Carlo Crivelli, 1486)

Location: National Gallery, London, UK

Image Citation: "The Annunciation," Wikipedia, available at: https://en.m.wikipedia.org/wiki/
The_Annunciation,_with_Saint_Emidius (Public Domain).
5. "The Crucifixion" (Fresco, 1350)
Location: Monastery of Visoki Dečani, Kosovo
Image Citation: "The Crucifixion (Visoki Dečani Monastery)," Wikimedia Commons, available at:
https://commons.m.wikimedia.org/wiki/
File:Crucifixion_of_Christ_-_Visoki_De%C4%8Dani_Monastery.jpg (Public Domain).

Government and Declassified Documents:

1. U.S. Government (1980). "Declassified CIA Documents on Remote Viewing". Retrieved from the National Security Archives.
2. U.S. Department of Defense. (2006). "DARPA Research on Quantum Computing and Light-Based Information Encoding".

Quotes and Creative Influence:

1. Nicks, S. (2011). "In Your Dreams" [Album]. Reprise Records. Quote on inspiration: "When you're being channeled, you just have to get out of your own way."

Archaeological and Fringe Research:

1. Hancock, G. (1995). "Fingerprints of the Gods: The Evidence of Earth's Lost Civilization". Crown Publishing.
2. Bauval, R., & Gilbert, A. (1994). "The Orion Mystery: Unlocking the Secrets of the Pyramids". Crown Publishing.

Meditation and Biofield Research:

1. Davidson, R. J., & Lutz, A. (2008). "Buddha's Brain: Neuroplasticity and Meditation". "Scientific American Mind, 19" (2), 48-53.
2. Schwartz, G. E., & Russek, L. G. (1997). "The Living Energy Universe: A Fundamental Discovery That Transforms Science and Medicine". Hampton Roads Publishing.

Additional Resources:

1. National Geographic. (2002). "The Secrets of the Pyramids".

National Geographic Magazine.

2. Tesla, N. (1919).

"My Inventions: The Autobiography of Nikola Tesla". Retrieved from (https://www.teslasociety.com/biography.htm)

3. OpenAI. (2024). ChatGPT (GPT-4) Computer software. https://openai.com/chatgpt

While ChatGPT provided valuable assistance with proofreading, all final decisions regarding content, structure, and style were made by the author.

4. Lambda Wave GIF on page 17 of Chapter 1: This file is licensed under the Creative Commons Attribution-Share Alike 4.0 International license. https://commons.m.wikimedia.org/wiki/File:1D-Wave.gif

5. Stellarium Labs S.R.L. (2024). Stellarium Mobile Plus (Version 1.12.9) [Mobile application]. Stellarium Labs S.R.L. https://stellarium-labs.com

To the Purple in My Tabernacle: Miss Liria

To L,

As I have progressed in my journey of life on earth, and my maturing through the biological aging process, spiritual aging process, mental aging process, and emotional aging process, I have become blind-less, deaf-less, mute-less, and numb-less. I now comprehend the difficulties you experienced when I luckily lived with you. The countless times you would rotate touching each finger with your thumb, searching for "it," and those that harass others through the medium of frequencies and vibrations. The countless times I was awoken in the night by your lack of physical and spiritual presence, only to find you at the marble table in 20FG, visually battling the harassment of evil imposed upon your face and body. At that time I had not matured enough to fully comprehend your battle, which is mutually experienced by every human being on this planet. Some are deaf, blind, and mute, therefore, do not experience or comprehend the battles they face upon biological death, and spiritual judgment. Those that are deaf, blind, and mute in this medium are purposely made so, to ensure when they experience the separation of soul and body they are imprisoned and hindered from proper transcendence. Your pain, agony, stress, and dismay during *times of being harassed may seem like misfortune, a curse, or Hades. I assure you that, although very difficult, it is not the aforementioned. Our biological forms are vessels for our spirits and souls. At times our biological forms must endure difficulties, which by design are intended to do so, to protect our spirits from the forces of evil and negative energies. You experienced, and still experience, the battle that one's spirit fights upon death of its biological form. However, you fight this along with billions of us, while both spirit and body are amalgamated. Those that do so have the lightest of hearts, allowing proper and heavenly transcendence upon separation of body and spirit. As you now know, I have experienced my body and soul beginning to sunder. However, the strength both my body and soul expresses <u>forcefully</u> overcomes the separation of the two. This is a result of repentance, the battles <u>you</u> also experience while biologically living, and the journey of my body and soul lacking completion of their unified purpose. Every human being has this duality of biological form and spirit. It echoes the age or æon with which we precessionally reside in at this time. Although in today's world many experiences, for both the blind + blind-less, deaf + deaf-less, mute + mute-less, can be falsified or simulated through technological means, the experiences I have had throughout my <u>entire</u> life provide a clear depiction and understanding of what is real and what is fake. Knowing the difference between the imposed, falsified, and simulated harassment, and the true, real, existential experiences, I can assure you that if you* *take the necessary steps to deviate the two that you will not only find solace and comfort, but also find the true memories or thoughts that should be contemplated upon separation of biological form and spirit. Therefore, I am certain some or many of your experiences that are, and were, similar to what I have personally seen were simulated and forced by individuals or organizations seeking emotional and spiritual energies, exuded by you, for their own benefit. I have been shown and told this directly and clearly. I wish I was older, more mature both spiritually and mentally, when we coexisted physically. Hopefully, that was not our only point in time when we do coexist physically. In the unfortunate circumstance that our physical coexistence has passed, I want to, of course, provide additional methods and techniques for your battles with the forces of evil: both simulated and real. My true love for you, Liria, as well as: the support and age-old teachings from the Roman Catholic Church, various spiritual and intellectual individuals, the guidance and direction from our beautiful and powerful military, those that have and do love me, and "We the People" as a collective entity, provide the following words of wisdom and advice for your frontline battles:*

"We win individually, yet triumph over all unifyingly."

When your fingers and thumb gravitate towards each other, refrain. This exposes the external influences applied to you.

Once exposed, during emotional (or "emotional") distress, you can begin truly battling the external forces applied to you. This can be done by opening your hands and ensuring your fingers are spread apart and there is no connection that can form a resonant conductive feedback loop to your harasser or entity feasting on your strong emotions. Universally, energy in all forms is an aspect of currency, albeit a universal currency that is utilized on a cosmic scale. Random movements, similar to shaking one's leg, tapping hands and feet, or any subconscious movement that is not advertent: it is your guardian (in religious beliefs) angel assisting you and maintaining your lack of knowledge at the same time. They're beautiful, intelligent, selfless entities acting and reacting for only <u>ONE</u> soul: yours. Regardless of religion they exist. I promise. Their hindering of situational knowledge isn't perverse, it is meant to protect your soul from acknowledging the evils that are applied to you. However, eventually it will need your cooperation and strengths your soul has achieved in this specific time and era to combat the evolving methods that all forms of evil utilizes to use your bodily form, and the emotions exuded from it, to their own benefit. When your Guardian Angel has an acknowledged and direct line of communication with your mind, body, and soul, it is <u>unbelievably</u> profound. My near-death experience, and experience with the bodily death process, showed me this since age 11. There is nothing like it, regardless of experiences in life. Trust me, I'm young, but very experienced with the fruits, vegetables, *meats, grains, and devious natures of life—I did date and live with you for years, he-he-he...*

Once you have opened your hands and welcomed your angel's presence; and yes, you have one and it isn't a fallen angel, the real battle begins. The days of checking your fingers, experiencing inadvertent facial expressions, needing to utilize substances to block "it", or playing/feeding into it to survive, will cease to exist. As the odd saying goes: 'You can see the devil dance in idle hands...' "It" is a competitive adversary. I won't deny that. But with the help of your Guardian Angel, your loved ones in life, and your desire for eternal salvation upon biological death, it is easily defeated. I promise that Liria. Repentance is key now for what is said in your mind when truly repenting is heard by three entities: your soul, your Guardian Angel, and the warden of Hades (Satan, Anubis, Abaddon, Apollyon, Iblis, Mara, Set, Demiurge...to name a few monikors...)

Ensure before you repent that there aren't external, synthesized, simulated, false individuals acting as the warden and eternal prisoner of Hades. When experiencing specific scenarios I have already mentioned, promptly put on headphones, or blast the following into your ears... I did this in my car with twelve speakers. It allowed me to move my head around and experience different binaural beats depending on the location of the speakers and the frequencies playing out of each one. In the car setting, where sound is confined and bounces off many objects, I played four different frequencies between 2,000 Hz and 18,000

Hz. This created an interference in the brain and coincidentally the overall output analyzed by external influences or monitoring by <u>humans</u>, was interrupted, resulting in a diminished application of bodily movements, facial expressions, and hand gestures. I was flabbergasted, Liria!! Due to the localized nature of the sound waves, and the ability to move around the cabin of my car, I was able to impede on both the external readings of human intervention, the internal effects of external influence, while my true Guardian Angel and I sat there realizing it all. This separated <u>ALL</u> external influences, and left us alone. At that point in time, we both acknowledged you, Liria. I immediately thought of you, Liria. I used that car to repeat those methods, both at high speeds and stationary. It worked. I migrated to headphones, ensuring I did not even know the true frequency, utilizing my lack of 20/20 vision to blur my phone's screen. Liria, this not only worked but caused an extreme reaction when I turned it off or stopped. Both I and my Guardian Angel laughed at their newly found vulnerability. I played frequencies in my car and headphones for months, it worked. The 'bath' of frequencies we reside in today was defeated by a localized bath of frequencies both surrounding my body and influencing my mind. In the first chapter of my book, <u>The Universal Language of Divine Light and Frequencies</u>, I discuss the mathematical equations involved when deciphering or utilizing this Information. When using headphones, play a frequency in one ear and a different *one in the other. Make sure you can't see or don't know the frequency played in either ear. This is important. It allows you to experience independence from binaural beats without allowing external influences to, at first, tap into you through frequencies. I guarantee they will attempt to force your eyes to view the specific frequencies. Once you know, and therefore they know, alter it. This creates a disturbance, even though brief, within the mixing process of modulated frequencies of the brain. If you need further explanation of this aspect, refer to chapter one of my book. It contains equations and examples of this.*

Now that you have deviated yourself from the forces of evil and can identify your own actions compared to the external input applied to you, the battle deepens but also becomes easier to fight. I am certain that when you were young and placed yourself in what you called 'La-La land,' you were inadvertently altering the 'f_difference' frequency by providing a change in the values expressed by your brain waves. At that age, you did not fully understand this, but you experienced a change in the negative energies applied to you. It was genius, which simply expressed your overall intellect and proper analysis of both mind and spirit. You are a genius, although odd and strange, a genius. Weird > Normal. At this point, return to 'La-La land,' but don't simply reside there, stagnant visually, bounce around the visual experience while providing all senses within the experience: auditory, olfactory, gustatory, somatosensory, and of course visually. Mentally providing altered values for your five senses in 'La-La Land,' or in technical terms your 'Sensorium' / Immersive Cognition, causes further disruption when external forces are both reading brain activity and actively altering it. As I'm sure you know, and have known since you were a child, this takes you from the marble table in 20FG to a different place, all while never physically moving. These values of sensory are reflected when monitoring brain activity. Essentially, if one were to analyze where Liria is, at that point in time, through brain waves, frequencies, and vibrations, their analysis would express 'La-La Land.' However, there must be a forever-changing and rapid occurrence of 'Nomadic Tranquility.' This is a state of inner peace or calmness despite constant movement or a lack of a fixed/permanent place of residence. Meaning, in this context, stay peaceful and calm, while traversing into 'La-La Land,' rapidly and continuously altering the environment of 'La-La Land' through all senses. I did this, and those that were reading my brain activity, in an attempt to surveil me both sinfully and furtively, saw your marble table in 20FG. I used that space, along with many other environments, as part of my own 'La-La Land' to ultimately cause a disruption in their values. They were both aware of this, and aware of my metaphorical message about you. I was directed to join the National Guard shortly afterwards...the National Guard is embodied with beautiful souls and beings, I promise you. When altering the environment of your 'La-La Land,' think of all types of environments rapidly. For example, go *from a beach, to the desert, to the jungle, to the city, back to the beach, then to the arctic tundra, then into the ocean, up into outer space, and then land at the marble table in 20FG. Not just visually, do this while providing values for all five-to-six senses. The sixth sense being <u>balance</u>. Don't just balance yourself in a physical sense, balance yourself spiritually and emotionally. This is key in maintaining your traversing battle or journey through the medium. You will undoubtedly experience words, environments, feelings, emotions, and all aspects of sensory that differs from your personal decisions when traversing the medium. We will record the "f_difference" in this situation, to ensure separation of yourself and external influences. Know that you are not alone in these experiences, and that I will personally and lovingly be there for you every chance I get. Hopefully distance doesn't impede on my assistance too often, or too greatly.*

By now you may feel confused, tired, and emotional, this is normal, and so is everything you have read so far. During the pandemic, sometime around May or June, we spoke over the phone for a short time. You were single, or about to be, so you answered immediately and were more open to conversing. It was lovely. You mentioned how you found yourself "talking to yourself" consistently while isolated like the rest of the world. I told you that this was a normal reaction the mind has when kept in isolation. This is true, and was what you were experiencing most of the time. However, it wasn't just your mind reacting or straying, it was also many people, entities, and beings not only communicating with you, but lightening your heart and keeping you company. At times the words said and the thoughts that occurred within you may have been difficult to process. At times it was difficult for me to process as well. But oddly enough, I could suddenly see and hear you, Liria. It was strange then, it was new to me at the time, and I wasn't sure if it was a result of the gigantic welt I found on my leg when leaving 20FG in October of 2019, if it was a result of my consumption of Malarone (Hydroxychloroquine) and Azithromycin two months prior, or potential COVID infection that I wasn't testing positive for. Yes, they prescribed me "that" cocktail of medication after seeing you in late 2019, staying at your home for one last night, at least one last night for now. Once I realized I could share the experience of almost all senses with you, many others, and eventually anyone I wanted to, I started my journey helping, healing, guarding, and at times interrogating anyone who dared to harass me or my loved ones. You are a loved one, yet no one is free from interrogation, when considering the weighing of the heart upon death can banish the soul to Hades indefinitely. I have heard awful, terrible, downright evil things that people have done. Thankfully, you are not included in that group. You have endured some harsh, awful, evil experiences throughout life. Your body has been used to perform actions that neither your mind nor spirit can recollect. How you were treated way back in 2012, when enduring and experiencing such evils imposed upon you, was barbaric, stupid, pathetic, unwarranted, unjust, terrible, and the definition of evil. Your personality, heart, and spirit deserved and still deserve better, I know this for a fact. Over the years I have ensured this statement is true and pure. Although for years I periodically experienced anger or dismay when recollecting the welt I had on my left leg after staying with you overnight in October of 2019, my experience with doctors refusing to even look or analyze my welt eventually exposed the truth to me. This no longer could overshadow your true soul and personality, which at times was hijacked by others in the eyes of darkness and evil. It is shockingly easy for many people to fall victim to this, but as technology has advanced, and the "bath" of frequencies that surrounds us in the medium has changed dramatically, it has inevitably become easier to differentiate the true mind and soul from external influences. These external influences can impede on our free will, very easily. But once this is recognized it frees the souls of the abused and used. I promise. These methods are used by the military on a regular basis. Marines, Navy Seals, Special Ops, Black Ops, and Green Berets are utilized in this manner to perform Top Secret duties in extremely sensitive situations. It ensures the soldier, or warrior, does not recollect the performance of the mission. This protects the warrior from interrogation by other entities, maintains a light heart for the warrior's judgment upon death, all while protecting and serving all of us. These methods have *unfortunately leaked into publicly accessible platforms, which has inevitably fallen into the hands of evil enemies and entities. But the double-edged sword of the modern bath of frequencies provides new weaponry against the forces of evil you experienced years before my presence, and unfortunately still do to this day. As we say: Stay Frosty! We are here for you. Jesus Christ held this double-edged sword, it is essentially the same as it was back then. It is why the crusades lasted over a millennia. Not because of the foolish belief systems in the mind of the person, and the governing powers of the world attempting to control the citizens, but because there was, and still is, this ongoing battle within those that are blind-less, deaf-less, numb-less, and strong. We aren't easily controlled, but at times we are controlled against our own free will. This is now recognized by civilians, soldiers, warriors, government employees, vigilantes, scientists, computer science engineers and technicians, and even terrorist organizations. We are all pissed the F*** off now. I, AJ, can see the color of your soul, Liria, and now so can the rest of us...Beautiful and vibrant Amethyst, Purple, Indigo, or Violet. A key element in the tabernacle intertwined with Cherubim, the second most powerful angel in the hierarchy of holiness. If you need to comprehend what my, and three-billion individual's angels are, refer to Chapter 7 of the book I have transcribed, titled: "<u>The Universal Language of Divine Light and Frequencies</u>." Or simply look above the Madonna in the painting titled: "<u>The Madonna with Saint Giovannino</u>."*

in the Palazzo Vecchio. That is what the Roman Catholic Church, and coincidentally, three-billion of us, consider to be our angels. They traverse the cosmos, our medium on this planet, and even our minds. They guide us, protect us, and are with us for the length of our biological-form's existence. I commanded, many years ago, an instruction for my guardian angel to aid you in your journey of life while I was not present. I truthfully and existentially feel the lack of its presence. It is still instructed to do so. When it returns to my biological form, and soul that resides in my form, it wreaks havoc, and reaps the evils attempting to destroy my body and soul. It is so powerful, I researched for years just to deduce what I am experiencing. In both language and mathematics, it is real. Just remember, I will always have it return to me, so be prepared to fight the battle without it at times. But I love you dearly, so it won't be often. Do not confuse this with the harassment and control of humanistic external influences. They play a role, and are typically always there, but feel so different from other entities that it is sometimes laughable. I have legitimately laughed at their presence. It is why AI will have a hard time triumphing over anything. My years of experience proving AI responses wrong, and breaking its algorithms, showed me this without the need for divine intervention. Divine intervention solidified this thought process. I have already spoken about controlling and hindering the application of forceful hand gestures, bodily movements, reactions, and even mindsets. However, there is more to this aspect. I have directly shown you the gesture that Jesus Christ utilized after multiple decades of experience. It is the "ICXC" hand gesture. This gesture cancels out the forceful impacts of other gestures. This includes, but it isn't limited to, the index-to-thumb gesture. It combats all other iterations of connecting one's fingers to both force exuded energy and control the individual. This can feel euphoric, enlightening, and can mimic positive feelings, which is confusing and misleading. Refrain from this feeling, it is not only the definition of fleeting but it is where evil mindsets and devious actions reside. It is meant to entice the person, just like all fleeting and short-lived euphorical feelings are meant to do. Like I previously stated, first start by opening your hands and extending your fingers, ensuring they are stretched and away from each other. Do this when you are angry, sad, emotional, contemplating thought, or generally clenching your fingers or fists. Monitor your movements, making the definite decision that you refuse to clench your hands and fingers. You may begin to observe your fingers twitch, start to clench, or migrate towards each other. This is you witnessing true free-will battling the forces of evil people, entities, organizations, and cults. This is why you experience bouts of lethargy, sadness, lack of energy, or depression. They have exhausted and expelled all of your serotonin, oxytocin, dopamine, norepinephrine, and epinephrine. These can be depleted when using narcotics and prescription drugs. Although certain narcotics can help with production of the aforementioned release of neurotransmitters, and although this is understood as part of the drug's mechanism of action to provide relief or euphoria, it is the brainwaves one suddenly resides in, and the individuals, entities, or external forces that utilize your body's experience of feeling and emotions that should be considered seriously. When experiencing anger, frustration, sadness, aggression, and any other emotion that produces the following hormones, neuropeptides, neurotransmitters, and chemicals that are controlled and utilized by multiple external forces, please flag this by both acknowledgment and reactive actions such as opening your hands and refraining from any hand gesture. Not only do you need this to occur, but so do we. Again, we are here for you, so be there for yourself: Serotonin, Dopamine, Norepinephrine, Epinephrine, Oxytocin, Gamma-Aminobutyric Acid, Endorphins, Glutamate, Acetylcholine, Cortisol, Melatonin, Histamine, Prolactin, Anandamide, Substance P, Phenylethylamine, Vasopressin, Endocannabinoids, Neuropeptide Y, Dynorphin, Enkephalins, Corticotropin-Releasing Hormone, T3 and T4 thyroid hormones, pregnenolone, beta-endorphins, neuropeptides, tachykinins, growth hormones, prostaglandins, histamines, leptin, ghrelin, estrogen, testosterone, progesterone, cholecystokinin, galanin, allopregnanolone, insulin, orexin, neuropeptide S, aldosterone, adiponectin, nitric oxide, and of course, adrenaline. These are the conduits to specific frequencies that are not only descriptive of all biological currencies, at least according to Homo sapiens, but the frequencies, vibrations, and the majority of chemical compounds that dictate our mental, emotional, physical, and spiritual states. It is more than just chemical compounds, reactions, and physical or spiritual states; it is a signature or watermark of those that are trapped and enslaved within a specific frequency, vibration, or the simpler term, a "neurochemical modulation or signaling." They reside within these frequencies, feasting and evidently surviving on the emotions and feelings exuded while riding these waves. This was unbeknownst to me, while both living in 20FG with you, and enjoying the fruits of life in my 20s. However, I am now both wiser and more experienced in this field, utilizing my daily occupations to assist anyone that beckons help with their trauma. I have not only been taught, but have both independently and intuitively learned these teachings, immortalizing what I have witnessed, assessed, and been taught, into my DNA, spirit, and relatives connected with me for an eternity, so be nice, my dear.

It is now time to discuss my niece's experiences, which are shared experiences with me through the medium of frequencies, vibrations, and the connection that microchimerism provides when utilizing the aforementioned. They have drawn pictures, expressed verbally, and have experienced all senses mutually with my unfortunate trauma and abuse. They were 3 years old and 6 years old at the time when I was first made aware of this.

Kids are innocent, and are not aware of what is normal, correct, or morally just. They were at an age directly between your horrific and awful experiences in the 1970s. It is an age that allows minimal memory or recollection, but utilizes almost every one of the forty-four previously mentioned "neurochemical modulations or signaling." I would wake up in the morning for years, up until meeting with you, Liria, in October of 2024, to sexual abuse from a distance. My nieces would cry and scream during these horrid moments. I remained celibate during this time, and aside from your light kiss I have remained so. It wasn't just my own sexual abuse that sparked my celibacy, but my two nieces' horrors they experienced almost daily maintained my purity. Their horrors coincided with mine, and I performed checks and balances to confirm this. Although my nieces aren't as intellectual as you were at their age, they made sure it was known that something wasn't right. Giada has, and still did up until our meeting in October of 2024, cried and screamed at various hours in the morning when I would experience sexual harassment at the same time while asleep and unconscious. I ensured this, by staying over their house only to wake up to their screams and terrors, finding myself officially raped or sexually abused. I initially thought I was causing their outcry, by verbally or mentally denouncing their actions (evils) applied to me while unconscious or asleep. I was incorrect, they woke me up to their cries. Giada cries more than Lucia, but Lucia was the one that you kissed and met personally. Thank you for what I observe as defense for her, whether inadvertent or not. Your experiences at their age, both horrific and unforgotten, echoed and assisted my family's genome. Your pain was not wasted, ill-acknowledged, or forgotten. Time, which is a bastard in some ways, doesn't defeat any of us this time. Expressed many times in this scripture written to you, Liria. It is why I say: "Leave the term 'age' to the Æons," for anything below such a magnitude is weak, irrelevant, and beneath all of us. You can thank my nieces for their innocence, my indestructible and persistent love for you and what you have endured, and the intelligence of my ancestors, our swamp-less intelligence agents and soldiers, and our inevitable and purposeful eternal existence in both the medium and reality with which we reside. I maintained my total celibacy for various reasons: to hinder the horrors my nieces experienced as a direct result of my potential actions if I wasn't celibate, both autonomously and jointly, and to indicate –properly– the external influences applied to myself and them. Performing this sacrifice was a pivotal point in time and aspect when being monitored by our armed forces and agencies that performed, and still perform, surveillance for security clearance purposes. It showed them what law enforcement agencies had overlooked for so many years. Just ask Epstein and P-Diddy, they exude what has been said. Although what you experienced in the 1970s wasn't stateside at the time, what you endured while a citizen here coincides with what is brought to light today. It isn't forgotten or ignored, simply delayed in terms of its application of judgment; your service to this country, both advertent and inadvertent, subconsciously and consciously, was never and will never be ignored or forgotten. The evils you endured were expressed to me, and exposed by me, AJ DelVecchio, and were not ignored. I'm sorry, I did not know you at age 3, 4, 5, 6, or 7 and on. The wrath I would have expressed would have killed others, especially if I was in the current field I am in. The beautiful resolution to this was immediately experienced by me: after seeing you and receiving the lightest of expressions of love...a simple peck on the lips, eliminated my horrors, eliminated the Bell's Palsy I experienced as well as the numbness in the mornings, and eliminated the other symptoms I endured since February of 2022 and at times earlier. Especially during the pandemic between December 2019 and February 2022.

In February of 2022 I was, by then, heckled at, and badgered about, a type of marijuana a manager at BMW, named Mike Dady, and a peon named Anthony Dominicus, expressed and tried to convince me to take for about 6 weeks. Their attitudes and personalities started to decline with me, and others, over the 6-week period. Viewing my job as in jeopardy, I asked for it but refused to pay for it. I immediately drove to my nieces' house, where my entire immediate family met me. I gave them each a portion, and explained to them my feeling of concern and overall distress about the situation and what was given to me. The point of me providing a portion of the narcotic *to each of them was to provide them with the evidence containing a potential and most likely lethal dose of a mysterious narcotic. I left, and went to see my friend well-versed in marijuana: Nic Ornaf. I asked him to check and analyze what was given to me. He looked at me, called me paranoid, and said it smelled fine. He offered to trade with me, or smoke it first to test it. He decided to smoke it first, so I let him, under his own will and apparent knowledge of the topic I knew minimal information about, to proceed with testing out Mike Dady's and Anthony Dominicus' cocktail disguised as simple marijuana. He seemed alive and coherent ten minutes later. I decided that the best course of action would be to pretend to partake in what was deviously and stupidly provided to me. I pulled on the smoking apparatus (bong), but halted when the smoke reached the mouthpiece. I wanted to read the reaction of Nic, making him think I inhaled. He turned and said "you fool," which is a biblical reference, while I showed him a full bong. He stopped talking. Since then, I can control his mouth with mine, I can pet his dog from a distance, I can visually see the changes this has applied to him. I was in the same room as him when he exhaled this smoke. However, my father also smoked <u>ALL</u> of what I gave him. I heard him for hours crying my mother's name, which was both heartbreaking and shocking. For months, he would tell me that he is in "5-D," which you and I both know is an impossibility for our biological forms and spirits at this time. This occurred on February 22nd, 2022...I woke up* *the next morning and showered before work. I'm not lying or being egotistical when I say: My eyes unnaturally stared at my crotch while a voice that was imposed and expressed to me said: "You can do that?!?" referencing how I dried my genitalia off by swinging it around in circles, something I had always done. The auditory experience ended with a shocked and imposed facial expression exclaiming "that is a huge d***." This is, and will never be considered, a complimentary or prideful experience in my life. It is one of the worst experiences in my life. This type of harassment occurred a few more times before I decided it was best to always shower in a bathing suit. I still do most of the time. After my nieces expressed their dismay and fea, as well, directly stating verbally: "I hate changing my clothes now Uncle AJ," I officially and whole-heartedly understood their pain. Being abused, and inevitably being measured by nearby objects over and over, my physical appearance and endowment will never be personally considered a curse, but rather a key aspect in gravitating the forces of evil, the pathetic, the underequipped, and the jealous individuals responsible for my niece's pain and terror. My sexual abuse, and most likely your experiences at your age, and-hopefully not, but most likely—other experiences you've encountered throughout your life will no longer be tolerated by society. This was a joint effort against evil that started before you were born and continued through your childhood, but dies with me. <u>F*** them</u>. The sinful nature of jealousy caused their destruction yet again. I am a warrior and a member* *of a team, every soldier is. When operating as a unit, the sad and sinful individuals we battle are disintegrated over time, with the help of the bath of frequencies we reside in, ironically meant to hinder us and enslave us...the beauty of irony in this situation really is profound. Even though your experiences when feeling lethargic, being glued to your bed or couch, at times recollecting memories that make you exude emotions utilized and addictively absorbed by the forces of evil in all iterations, may differ from sexual abuse at times, but they are from the same source. Welcome to the light side, which overpowers the darkness and evil. Naturally and inevitably, their existence resides in the pockets and <u>empty</u> space of nothingness, for these entities are <u>nothing</u> compared to us. Expected though. Light not only illuminates, but it exists in a timeless superposition that exceeds all other forces in terms of speed, information transfer, and co-existence of duality expressed to us everyday. For darkness without light doesn't comprehend true power, while light overpowers the darkness, which in turn allows to overcome all darkness. This duality of the light and our light, echoes the teachings and scriptures of a multi-millenia long existence of humanity. No one fucks with us and survives for long. Eventually the darkness gets absorbed by use, at that time useless and empty space, into a singularity that contains our light to hinder darkness from salvation and transcendence. Which is deserved.*

Light and its duality is a fierce and overpowering force naturally equipped to overcome darkness and triumph over its lack of illuminating life, creating life, maintaining life, and providing growth for life and everything it reaches. Even when round objects try to block and impede on the path of light, light figured out a way to not only use these round and molten objects but to traverse around them to allow a lack of impedance, allow the hindering of their obstruction (which such obstructions are destined to be futile) while utilizing these obstructions for the successful transmission of light and its information. Although light is obstructed by objects, it is only obstructed when viewed in specific areas of darkness. The light still reaches everywhere else, rendering the object useless when attempting to block light in its entirety. But what about "black" holes, one might say? They are anything but "black". They are only considered black due to the limitations of visible light. The information, frequencies, and non-visible (at least to us) light spectrums or frequencies emitted from the Einstein-Rosen bridges we call "black holes" are extremely intense, allowing light to yet again find another passageway away from the darkness to transmit its illumination to its destined and required destinations. Although the light that is contained by an Einstein-Rosen bridge appears to not escape, it is actually deposited in a different time and space from its initial absorption by the event horizon. This is the opposite of total destruction. This is an example of what it expresses through its waveform when viewing its interference pattern. It is everywhere until an "observer" (hahaha) tries to pinpoint its specific location at that specific time in a specific space. How foolish?!? To think that darkness can both mimic light and destroy it. It is the best joke the cosmos in all of existence can express. This is both metaphorical and literal. Another beautiful example of duality expressed by the divinity of light in all of its forms. This light shines upon me frequently but not always. It ensures that when something is severely incorrect with either me or those in my life that assist me on my journey, it does not allow the obstruction to block my pathway to living life, creating life, maintaining life, and administering both it's and my message. It has methods of doing so that are still unbeknownst to science, experts, and professionals. It is the beauty of duality in both the literal and metaphorical forms, and until the end of time itself, it can never be truly deciphered or overcome in its entirety.

Love,

The Universal Language
of
Divine Light and Frequencies

A Theory by
A.A. DelVecchio

Ephesians 4 8-11
Revelation 1 17-19

HEAVEN

The Universal Language
of Divine Light and Frequencies
By RJ DelVecchio

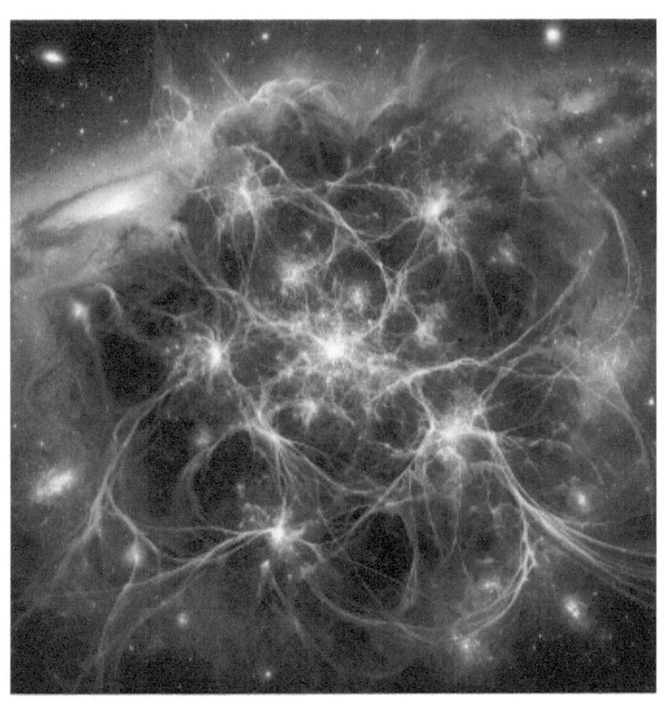

The Universal Language
of
Divine Light and Frequencies
by A.J. DelVecchio

About the Author

Anthony J. DelVecchio was born and raised in the United States of America. Through his persistent curiosity about science, physics, space, spirituality, and divine purpose, he studied for decades. His parents consistently nurtured his inquisitiveness, both answering his questions and encouraging deeper contemplation. Many individuals throughout his life, whether through love or friendship, contributed to expanding his perspectives, each offering some insight into his constant questioning. With a desire to understand and engage with the world or medium in which we all reside, through an absence of hate and an abundance of love, he was able to transcribe The Universal Language of Divine Light and Frequencies and the series The Universal Theory. Hopefully, his inspirational curiosity, disciplined determination, open-mindedness, and both parent's ability to raise a person with love and endearment will transcend to those who read his literature, assisting them in the positive progression of their lives.

A.J. is a member of the United States Army National Guard. He is a devoted Roman Catholic, with multiple near-death and out-of-body experiences related to the body's death process. He is a stoic, empathetic, calculative, and determined individual who forms opinions based solely on logic and truth. As a guardian of his nation, under oath he has committed to "We the People" and to protect and serve every member of the United States. He fulfills this duty physically, mentally, and spiritually through both action & literature. We hope his words are not just seen, but also read and heard between the lines he has transcribed for us.

A comment from those both within and outisde his family tree:

"...of the Ancient Ones living with us in the modern world..."

The surname "DelVecchio" is directly translated to: "of the old," "of the old tribe," or "of the ancient ones." It refers to the oldest living Hebrew-Italian descendants who are alive today. If you read A.J.'s books, you will understand the language that describes how cellular memory, DNA, and the medium around us immortalizes the experiences and knowledge of ancestry and ourselves to assist in guiding the lives of every lineage's future generations. It is why many centuries ago individuals within his lineage were provided this surname. A gift to all of us from our past. **ICXC**